"Smart meters are both a major weapon in the han⟨ ⟩ a subject of some public controversy. This important book explains what smart meters can do, how they are used in different countries, and how we can best take advantage of them to benefit consumers and promote decarbonisation."

—*Martin Cave, Professor, Chair of Ofgem,*
the UK energy regulator

"Smart meters are regarded as critical for the digitalisation of our electricity system, and yet there has been limited understanding and evidence about the value and benefits of these multi-billion-pound investment programmes. This book tackles its subject with genuine passion, showing in innovative ways how smart meters relate to our everyday lives and explaining the costs and benefits of smart meters in Europe and beyond. It is the one-stop book for anyone working or studying issues related to smart meters in industry, policy or research."

—*Jonathan Graham, Head of Markets and Strategy*
for Distributed Energy Systems, Siemens UK

"A long-awaited must-read (even must read twice), this book takes a comprehensive and innovative approach to electricity smart metering. Many thanks to Jacopo for opening the smarter meter blackbox from all possible angles and using his established legitimacy to explain to the reader all the alternative choices and possible outcomes".

—*Jean-Michel Glachant, Professor, Loyola de*
Palacio Professor in European energy policy and
Director Florence School of Regulation
(European University Institute)

"Flexible demand will be critical in future low carbon electricity systems, and yet understanding of how to achieve that remains inadequate. It is clear that smart meters will be needed. This book provides a timely and comprehensive assessment of the economics of that investment."

—*Nick Eyre, Professor, Director, Centre for*
Research into Energy Demand Solutions

"Understanding the responses of consumers in the smart, flexible and decarbonised energy system of the future is critical to its success. This book distils the findings of cutting-edge research on the economics of smart meters and hence represents a key element in understanding the role of costs in how people use energy services."

—*Neil Strachan, Professor, Director,*
UCL Energy Institute

"Drawing on an impressive range of interdisciplinary research, this book opens up economic, social and policy aspects of smart meters and energy demand. It makes an excellent read for anyone interested not only in smart meters, but also in the use of economic appraisals as part of large infrastructure projects."

—Susana Mourato, Professor, Head Of
Department of Geography & Environment,
London School of Economics

Appraising the Economics of Smart Meters

This book focuses on the economics of smart meters and is one of the first to present comprehensive evidence on the impacts, cost-benefits and risks associated with smart metering.

Throughout this volume, Jacopo Torriti integrates his findings from institutional cost-benefit analyses and smart metering trials in a range of European countries with key economic and social concepts and policy insights derived from almost ten years of research in this area. He explores the extent to which the benefits of smart meters outweigh the cost, and poses key questions including: which energy savings can be expected from the roll out of smart meters in households? Is Cost-Benefit Analysis an appropriate economic tool for assessing the impacts of smart metering rollouts? Can smart meters play a significant role in research on people's activities and the timing of energy demand? Torriti concludes by providing a much-needed survey of recent changes and expected future developments in this growing field.

This book will be of great interest to students and scholars of energy policy and demand and smart metering infrastructure.

Jacopo Torriti is a Professor of Energy Economics and Policy in the School of the Built Environment, University of Reading, UK. He is Co-Director of the Centre for Research into Energy Demand Solutions (CREDS) and Principal Investigator of projects funded by EPSRC on Residential Electricity Demand: Peaks, Sequences of Activities and Markov chains (REDPEAK) and Distributional Effects of Dynamic Pricing for Responsive Electricity Demand (DEEPRED). He has authored the books *Peak Energy Demand and Demand Side Response* (Routledge, 2015) and *Energy Fables: Challenging Ideas in the Energy Sector* (Routledge, 2019). He is a member of Ofgem Academic Advisory Panel and Defra Economic Advisory Panel. Before joining the University of Reading, Jacopo worked at the London School of Economics, the University of Surrey and the European University Institute. He obtained a PhD from King's College London, a Master from King's College London and a Laurea in Economics from Università di Milano.

Routledge Studies in Energy Policy

Business Battles in the US Energy Sector
Lessons for a Clean Energy Transition
Christian Downie

Guanxi and Local Green Development in China
The Role of Entrepreneurs and Local Leaders
Chunhong Sheng

Energy Policies and Climate Change in China
Actors, Implementation and Future Prospects
Han Lin

Energy Efficiency in Developing Countries
Policies and Programmes
Suzana Tavares da Silva and Gabriela Prata Dias

Ethics in Danish Energy Policy
*Edited by Finn Arler, Mogens Rüdiger, Karl Sperling,
Kristian Høyer Toft and Bo Poulsen*

Mainstreaming Solar Energy in Small, Tropical Islands
Cultural and Policy Implications
Kiron C. Neale

Appraising the Economics of Smart Meters
Costs and Benefits
Jacopo Torriti

For further details please visit the series page on the Routledge website:
http://www.routledge.com/books/series/RSIEP/

Appraising the Economics of Smart Meters

Costs and Benefits

Jacopo Torriti

Routledge
Taylor & Francis Group

LONDON AND NEW YORK

earthscan
from Routledge

First published 2020 by Routledge

2 Park Square, Milton Park, Abingdon, Oxon OX14 4RN
605 Third Avenue, New York, NY 10017

Routledge is an imprint of the Taylor & Francis Group, an informa business

First issued in paperback 2021

British Library Cataloguing-in-Publication Data
A catalogue record for this book is available from the British
Library

Library of Congress Cataloging-in-Publication Data
Names: Torriti, Jacopo, author.
Title: Appraising the economics of smart meters : costs and
benefits / Jacopo Torriti.
Description: First edition. | Abingdon, Oxon ; New York, NY :
Routledge, 2020. | Series: Routledge studies in energy policy |
Includes bibliographical references and index. |
Identifiers: LCCN 2019057254 (print) | LCCN 2019057255 (ebook)
Subjects: LCSH: Recording electric meters—Cost effectiveness. |
Electric power production—Costs. | Electric power consumption.
Classification: LCC TK393 .T67 2020 (print) |
LCC TK393 (ebook) | DDC 333.793/23—dc23
LC record available at https://lccn.loc.gov/2019057254
LC ebook record available at https://lccn.loc.gov/2019057255

ISBN: 978-0-367-20336-8 (hbk)
ISBN: 978-1-03-217316-0 (pbk)
DOI: 10.4324/9780367203375

Typeset in Times New Roman
by codeMantra

Contents

Figures

Tables

Preface and acknowledgements

This book is the outcome of research I have been carrying out on smart meters more or less since I completed my PhD at King's College London in 2007. Although smart metering is not the main field of my research, I find smart meters an attractive topic because they expose some of the potential of infrastructure we develop for a better future, the constraints of the markets and the organisations which roll out smart meters and the challenges of the society we live in. Smart meters are about measuring the electricity which runs through the wires in our buildings. They come with a promise of making known what was previously unknown, tangible what was previously intangible, flexible what was previously inflexible. By researching smart meters, questions emerge around the infrastructure we build, how we replace what was previously in place and how we envision a more automated future.

For those who operate in the energy market, smart meters are about opportunities to charge more for electricity consumption at certain times of the day, concerns around who will have to pay and opportunities to save money. For those interested in the common good, smart meters represent a complex investment decision, in which questions emerge around what the right price to pay is for higher interoperability with the smart grid and cybersecurity of consumers. For social scientists interested in behaviour and social practices, smart meters represent a yet underrated treasure which has the potential to uncover not only how individuals respond to direct feedback, but also how synchronised society is around certain times of the day.

This book frames smart meters in terms of their economic dimension. This is because the question of whether their benefits outweigh the costs has been challenged at different scales by a variety of actors, including media, consumers' campaigns and academia. In presenting evidence over the costs and benefits of smart meters, the book seeks to provide evidence on their economic advantages and disadvantages. I am aware that smart metering technology comes in a variety of forms with a number of different functionalities. They may also be connected to wider systems which for their ability to collate and distribute data at a higher frequency than ever before are often termed 'smart'. When combined with 'smart' elements within distribution and transmission networks, these wider systems are frequently referred to as 'smart grids'. This book refers to smart metering in relation to the key features that have the potential to (i) facilitate greater engagement from residential consumers in the retail electricity markets; (ii) reduce grid losses and operational costs; and (iii) reduce energy use and load-shifting. The book does not present new evidence on technological variation, although inevitably some communication, interoperability and cybersecurity features are discussed as they generate different cost figures. Instead, the book offers a platform for debating the role of smart meters in relation to its economic, policy and social dimensions. This is done by interrogating the role of economic appraisals in energy policy, reviewing empirical evidence on assumptions around behavioural change and highlighting smart metering functions in association with changes in pricing approaches.

The book commences by introducing recent changes in energy systems and the context in which smart meters are rolled out. Chapter 1 addresses the following questions: which role have smart meters been attributed in the decarbonisation of electricity demand; which issues have been raised following the European Commission recommendation on national roll-outs of smart meters? Chapter 1 presents the arguments which have been put forward in favour and against smart meters, hence making the case for the need for a book like this, presenting evidence around the economics of smart metering.

Smart meters measure consumption with a higher temporal granularity than previous meters. Which lessons can be learned from previous electricity meters; how did units measuring electricity demand change over the years; and which role did the emphasis around cost reduction play in determining the various typologies of meters that have been installed over time? These are questions that Chapter 2 seeks to address. This is informed by a classification of main cost categories for smart meters as well as categories of benefits.

From the experience of cost–benefit analysis in different European countries, which units of costs and benefits vary the most? Which contextual drivers lead to significant impacts for different market stakeholders? How are benefits relating to aggregate energy savings and load shifting computed in cost–benefit analyses? Chapter 3 presents estimates of costs and benefits of smart meter roll-outs in Germany, the United Kingdom, the Netherlands, Ireland, Hungary and France.

How is smart metering cost data typically collected and how is its price publicly procured? Which are the implications of capping smart metering market price and depreciating smart meter costs? Chapter 4 addresses these questions by presenting evidence on economic appraisals of smart meters in Romania based on pilot studies carried out by the Romanian Distribution System Operators.

Is cost–benefit analysis the right tool to assess the economic advantages and disadvantages not only of smart meters but also of energy policies in general? What is the role of cost–benefit analysis in decision-making in energy policy? How do different types of data and models shape variation in the Net Present Values within cost–benefit analysis? Chapter 5 explores the extent to which cost–benefit analysis, as is the case for national roll-outs of smart meters, adequately appraises a wide array of societal impacts. Cost–benefit analysis is confronted with other appraisal and evaluation tools which have been used to assess the impacts of energy policies and regulations.

What is the empirical evidence that smart meters prompt behavioural change, thanks to the real time feedback they display? Does the literature produce conclusive results in terms of smart meters and energy savings? Which factors matter in order to explain variations in net conservation effects? Have net conservation figures changed over time, and how does this relate to the reliability of the studies? Chapter 6 presents a meta-analysis of smart metering trials and conservation effects, focusing on sample sizes of the trials and sample selection methods.

Which conceptual views can enable a better understanding of smart metering and energy demand? How do behavioural theories and social practice theories relate to smart meters? Chapter 7 makes connection between smart meters and research on people's activities and the timing of energy demand, on the one hand, and key concepts of behavioural theories and social practice theories, on the other hand.

Which developments in electricity generation and demand will involve smart meters in the future? Will smart meters interact with appliances and devices in the home? Will they be integrated with various forms of energy as a service and blockchains? Which types of tariffs will be unlocked by the smart meter? Chapter 8 questions what the smart future may look like, the future role of smart meters and recapitulates the six key messages which this book offers.

I wrote this book mostly without the aid of my smart meter.

However, several researchers, colleagues and practitioners helped me with collecting, analysing and discussing the material which is presented in this book. Some of them work with me in the School of the Built Environment, University of Reading: Mate Lorincz, whose work is underpinning Chapter 7, and Timur Yunusov. Others work in different organisations. The book was written during my time as a Co-Director of the Centre for Research into Energy Demand Solutions (CREDS) funded by the UK Research and Innovation. CREDS has offered me invaluable opportunities to learn from some of the best UK researchers in the area of energy demand. It is a large centre and I will not have time to thank all individuals here, but I will make sure this happens in person. My thinking on the relationship between energy demand and social theories has been substantially influenced by ideas around social practices as part of the DEMAND centre. Other projects enabled me to do research on the technical and economic features of Demand Side Response: the EPSRC Fellowship, 'Residential Electricity Demand: Peaks, Sequences of Activities and Markov chains (REDPeAk)',

EP/P000630/1; the EPSRC project 'Distributional Effects of Dynamic Pricing for Responsive Electricity Demand (DEePRED)', EP/R000735/1; the EPSRC project 'Reshaping Energy Demand of Users by Communication Technology and Economic Incentives'; and the TSB project 'Assessing the Benefits of Demand Side Response Participation in a Capacity Market'. As part of these projects I learned from several colleagues and practitioners.

The work on Romania was made possible by a collaboration with the World Bank. I would like to thank staff of the World Bank and the Romanian Government for their time in explaining the complex dynamics of smart meters there. I also learned during my time as a member of the Public Interest Advisory Group on smart meters, in the UK which brought together a broad range of public interest stakeholders to consider how smart meter data can be put to best use to further public policy goals and aid in the energy transition and whether the data might be accessed by government and other organisations for public interest purposes while safeguarding consumers' interests, including on privacy.

While I was working on this book, I had the opportunity to present parts of the book at various conferences (European Energy Markets, International Association for Energy Economics, British Institute of Energy Economics), workshops and seminars organised by the University of Oxford, City University, UCL, Université Paris-Dauphine, University of Lancaster. The colleagues who contributed to my thinking at these events are many, and it would be too difficult to list them here.

The research underpinning this book was carried out while I was based at different universities. For this reason, I would like to thank Ragnar Löfstedt from the King's Centre for Risk Management at King's College London for his doctoral supervision; Pippo Ranci for training me on energy demand when I was at the Florence School of Regulation, at the European University Institute; Matthew Leach from the Centre for Environmental Strategy at the University of Surrey, who really initiated me on topics related to smart meters; and various colleagues from the London School of Economics for when I was teaching and doing research there.

On a more personal note, I would like to thank my family. My mother, who is a writer and has passed me the passion for reading and writing, my father and my sister who understand better than me some of the dynamics of markets and organisations.

This book would not have happened without the support of my wife. She is a constant source of intellectual inspiration because of how she challenges some of the mainstream thinking around economics and ensures I take some distance from it.

1 Are smart meters worth the cost?

Waiting for meter data

My second son has just re-discovered it. It has been away in a drawer in the kitchen along with other objects we use very seldom (knife sharpener, birthday candles, party cups, etc.). It had been lying there for many months, maybe even a whole year. My first son once played with it for 20 minutes or so and then returned it to that drawer. Now, this rarely used object is getting someone's attention once again – the attention of a four-year old.

- What is it *papà*?
- It's a smart meter. Well, an in-home display.
- What does it do?
- It's supposed to let us know how much electricity we are using. For instance, when the dishwasher is on, I look here and see some numbers.
- Wowww! Shall we do it?
- It doesn't really work, but you can press this button and see some zeros here.

I forgot that when we switch it on, we do not see the zeros, but a very standard message: 'waiting for meter data'. This was the first disappointment I had with my smart meter. There is nothing smart about having this in-home display at home and not being able to know in real-time how much electricity we are using and how much money we are spending because we live on the third floor of an apartment block and the in-home display cannot receive information from the smart meter. While my son starts pressing all the buttons the device can offer, I am thinking of other uses this might have. Years of research in this area prompted me to think that after all the benefits of the smart meter are not just about direct feedback on energy usage. Smart meters are also about opportunities to get involved in demand-side flexibility and participating in residential demand-side response, on which I wrote another book (Torriti, 2015). On the one hand, this sparked further frustration as I came to realise that in the future I will be cut out of opportunities to change retailers in real time, move across time of use tariffs, benefit from dynamic tariffs and one day – who knows – even have the smart meter communicating with a controller which minimises the costs of my electrical appliances as it selects when they are on and off based on the cheapest (and greenest) available tariffs. On the other hand, these unmaterialised opportunities make me hope that regardless of that 'waiting for meter data' message, progress is under way and the future might be brighter for others. For now, when it comes to meter readings, I can take my in-home display out of the drawer, out to the corridor, down the lift and – once on the ground floor – finally after a couple of minutes my 'not anymore in-home' display starts connecting to the smart meter. Basically, it is a substitute to the key of the communal door behind which all the meters of our block of apartments are.

My son finished playing with the 'smart meter'. The smartest thing he did was to put it back in the drawer. I start thinking that smart meters got a lot of bad press for different reasons but there is progress in what they set out to achieve. Writing a book on the economics of smart meters may not be the smartest thing to do, but it is what I decide to embark on.

Context

Over the past decade climate change imperatives have led governments of several countries around the world to move away from fossil fuel-based power generation and to increase shares of cleaner power generation technologies into power systems. This means that the generation of electricity is becoming less dependent on a few centralised power stations as these are being gradually replaced by a multitude of smaller plants. The highest level of decentralisation of power generation consists of industrial, commercial and residential buildings producing electricity from smaller units of renewable generation (Hanna et al., 2018). Concerns over security of supply feature among the explanations leading to policy decisions around the decentralisation of the power systems (Allen et al., 2008). Distributed power generation assets can improve the utilisation of local energy resources, offer greater resilience and potentially be more cost-effective (Chmutina & Goodier, 2014). The reduction in costs compared with more centralised systems is justified by the fact that generating power locally reduces some of the costs of transmission and distribution system upgrades, and the additional generation resources can contribute to grid reliability as the risk of technical failures is spread across a wider portfolio of generation.

Policies which have steered the way towards a higher integration of renewable energy are generally paired with the vision of a model where the power system is connected by a two-way grid and where resources such as physical storage, active demand management and the output of hundreds (if not thousands) of small power generation units will be balanced in real time with the aid of intercommunication links.

The integration of decentralised renewable sources of electricity also presents challenges as these are associated with an increase in the rigidity of power supply. The intermittent nature of their supply along with the lack of control over their outputs and their limited system inertia poses new challenges with regard to the balancing of demand and supply. Flexibility provision has historically been a responsibility of the supply side (Grünewald & Torriti, 2013). The transition to a net zero or low-carbon energy system will depend not only on the decarbonisation of power generation through an increased uptake of intermittent renewables, but also – and to a great extent – on the ability to find alternative ways of providing the flexibility needed to maintain the system balance (Barton et al., 2013).

It is against this background that smart meters are introduced as part of demand-side device to be integrated with the changing needs of electricity supply. In a nutshell and as way of introduction, smart meters are expected to provide a means of collecting high-resolution energy demand data at end-use point.

Decisions around large public investments in smart-metering infrastructure are very difficult for policy-makers without the support of strong empirical evidence. This book presents research on economic analysis underpinning decisions to invest in smart-metering infrastructure. Smart meters are extremely interesting because they are part of a larger plan to transform electricity systems, making them smarter, greener and richer in terms of data exchange. Smart meters are the first and/or last items of very large infrastructure and are also very close to the people. They are supposed to capture the rhythms of everyday life (at least with regard to electricity demand), measure levels of consumption which may change depending on price and time of day, and send signals to users warning about higher tariffs or peaks due to weather conditions, nudging customers to make them 'change behaviour'. The smart meters, with different degrees of human intervention, are also supposed to interact with the grid, sending data about demand which are of use for those who manage distribution networks, offering market opportunities for rewards based on temporary reductions in energy consumption. The information which is captured, measured and transmitted by smart meters is of interest not only to energy organisations (suppliers, aggregators, distribution system operators, transmission system operators), but also to non-energy organisations, which can derive from them useful information in terms of understanding occupancy patterns, consumption trends and market segmentation.

Overview of smart-metering roll-outs

For the past two decades there have been programmes and projects in place in various countries for the roll-out of smart meters. In 2009 the European Commission's Third Package for the liberalisation of energy markets marked a monumental change in smart–metering history as tens of countries commenced their roll-out plans in its aftermath. For this reason, two time periods are presented here: pre-2009 and post-2009.

Pre-2009: pioneering countries

Prior to 2009 smart meter projects had already taken place in Italy, Sweden, the Netherlands, Canada, Australia, California (USA) and Northern Ireland. The Italian utility ENEL introduced smart meters in 2001 as a result of an in-company investment decision. The rationale for this decision was based on expected savings or revenues in the areas of purchasing and logistics, field operations, customer services and revenue protection, which relates to a better understanding of where fraud occurs within low-voltage distribution networks. Over the years, I have been told by people working for ENEL that an economic business case existed, and this was not in the form of a traditional cost–benefit analysis (CBA). In fact, neither the Italian energy regulator nor government had any significant influence on requirements ENEL had to fulfil. This means that ENEL could choose the type of meter and the communication infrastructure independently and opted for a smart electricity meter that communicates through power-line communication (PLC) to the nearest substation (Torriti, 2012). By the end of 2005, ENEL had 27 million smart meters installed, of which 24 million meters being remotely managed and bimonthly read (Alberini et al., 2019). The early move by ENEL meant that the Italian regulator was confronted with challenges such as providing smart meters to the minority of consumers not covered by the ENEL and ripping the benefits of the investment made, which was estimated at about 2.1 billion Euros (Meeus & Saguan, 2011).

As a response to these challenges, the Italian regulator issued a mandatory roll-out of smart meters in 2006 and gradually increased operating costs in metering activities reaching the level of 7.1% at the end of the mandatory roll-out period (Lo Schiavo et al., 2013). As of today, more than 95% of low voltage customers, also those located outside ENEL areas, are provided with smart meters.

In Sweden some utility companies developed smart-metering pilot projects in the early 2000s and the government expected economic benefits in terms of energy savings. In 2003 the Swedish government issued a law which contained an obligation for monthly meter readings for all electricity users by 2009, hence creating the conditions for the introduction of smart meters. Since then, investments in smart metering have developed in a faster rate than required by law (Van Gerwen et al., 2006).

In the Netherlands the smart meter roll-out was planned since 2004 as a way to improve demand response at the level of small consumers. However, the regulated roll-out could only start eight years later due to unanticipated resistance from the public in terms of privacy rights infringements (Hoenkamp et al., 2011). An independent cost-benefit analysis was carried out as part of the decision-making process (see Chapter 3). This also followed the development of pilot projects (Van Gerwen et al., 2010).

In Ontario in Canada, increasing electricity demand peaks were the main factor for smart metering. The Ontario Energy Board installed about one million smart meters up to 2008 with plans to cover all 4.3 million Ontario customers by the end of 2010 (Torriti et al., 2010).

In Victoria in Australia, smart-metering installation started in 2006. The rationale consisted of increasing summer electricity demand peaks by air conditioning causing extra investments in low use plants and the linking of wholesale and retail markets.

Smart meters in California were introduced with the aim to increase the reliability of electricity supply in this state, through the reduction of consumer peak demand. California has a summer peak demand for power during approximately 50–100 hours per year. This peak is mainly due to the increasing use of air conditioners. The main energy agencies of California saw demand response as an important mechanism to decrease this peak.

2009: the EU directive common rules for the internal market in electricity directive (2009/72/EC)

The European Commission's Third Package for the liberalisation of energy markets recommended that member states should implement smart-metering roll-out plans which should achieve 80% penetration by 2020, if an economic assessment proves the implementation as feasible.

A European Commission interpretive note (pp. 8–9) for the Directive of 2009 states that:

> Where an economic assessment of the long-term costs and benefits has been made, at least 80% of those consumers who have been assessed positively, have to be equipped with intelligent metering systems for electricity by 2020 ... where no economic assessment of the long-term costs and benefits is made, at least 80% of all consumers have to be equipped with intelligent metering systems by 2020.[1]

Interpretations of how a consumer is 'positively affected' by smart metering (or a category of consumers, such as the fuel poor) vary, but the default option leans in favour of a high level of roll-out.

The rationale for the EU Directive originates from the fact that customers lack means for understanding their energy consumption and a limited number of incentives are in place to respond to changes in prices. A few factors undermine effective participation of the demand side in Europe: the lack of real-time price information reaching consumers; regulated retail prices in some countries; outdated metering technologies; system operators focused on supply side resources; and an approach traditionally averse to demand-side participation. Smart-metering implementation plans were developed under the European Commission's objectives of strengthening energy efficiency, mitigating greenhouse gas emissions and promoting renewable sources of energy. A wide roll-out of smart-metering technology was supposed to ideally create the platform for an informed consumer capable of responding to prompts from the supply side. Responding to changes in supply is critical in the electricity system because the production of power (generation) must always exactly match the usage (demand). As demand varies through the day and across the seasons, the system must be able to maintain the perfect balance of supply and demand.

Post-2009: laggard countries

CBAs provided positive outcomes, and electricity smart meters roll-outs were planned for in 16 member states: Austria, Denmark, Estonia, Finland, France, Greece, the Republic of Ireland, Italy, Luxemburg, Malta, the Netherlands, Poland, Romania, Spain, Sweden and GB. Seven member states (Belgium, the Czech Republic, Germany, Latvia, Lithuania, Portugal and Slovakia) found negative CBAs for the large-scale roll-out of smart meters. However, Germany, Latvia and Slovakia found positive CBAs for a subset of consumers. Some of these CBAs are reviewed in Chapter 3. As of 2014, three member states had already largely completed their roll-outs of electricity smart meters: Finland, Italy and Sweden (European Commission, 2014). As of 2019, Estonia, Malta, Spain and Denmark also completed a wide-scale roll-out for smart meters. Most countries are expected to reach at least 80% penetration of smart meters in the period between 2020 and 2025. About one third of the member states will roll out smart meters by 2030 or later, as their latest CBAs are still negative. Compared to the initial 2009 targets, fewer member states are in line to reach the deployment rate of at least 80% by 2020. Figure 1.1 shows the different timescales of smart meter roll-outs in EU countries. This shows that a minority of countries either completed the roll-out or has plans for 100% completion in future years. In essence, Figure 1.1 can be seen as the ultimate outcome of the CBAs which were performed by single EU countries.

Arguments in favour and against smart meters

Are smart meters an essential part of a zero-carbon future or a waste of money? With a few more nuanced positions and some undecided, the arguments in favour and against smart meters tend to be developed starting from radically different positions.

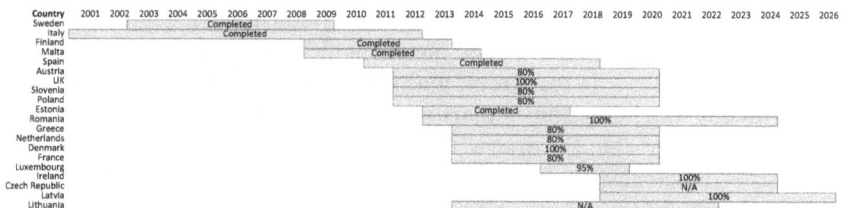

Figure 1.1 Timescales of smart meter roll-outs in EU countries.

Advocates of smart metering list several benefits, including lower metering cost, energy savings for residential customers, more reliability of supply, accurate bills, variable pricing schemes to attract new customers and easier detection of fraud. On the other hand, some argue that smart meters at the societal level may bring about more costs than benefits. For instance, there is limited evidence so far that smart meters will save energy (or money) to end users. The decision on whether and how to spend several hundreds of millions of Euros on such a radical change for electricity systems cannot be based solely on perceptions, lobbying and private interests. Moving beyond the positions of those who have a vested interest in the deployment of smart meters, the largest academic enthusiasm has been shared between technological optimism around the digitalisation of networks, the higher interconnectivity of the smart home with a modern smart grid and the economists' visions of demand-side flexibility, with smart meters playing a vital role in facilitating cost reductions from network reinforcements, consumers ripping returns in terms of lower bills at times of the day when electricity supply from renewables is plentiful and avoided costs of inefficient additional generation.

On the contrary, arguments against smart meters tend to emphasise increases in unnecessary costs, privacy and safety issues.

With regard to costs, this is one factor for the line of enquiry of this book. Some of the excessive costs arguments tend to be made within the context of national roll-outs. For this reason, this book presents a variety of national experiences and comparisons across countries as part of the analysis. For instance, in the United Kingdom, a cost which is not presented in the official impact assessments and is hardly mentioned in media debates about the economics of smart meters consists of the communication campaign to convince consumers about the advantages of smart meters. In recent years I have become very sceptical of campaigns in the United Kingdom which involve buses and grand slogans. The promotional campaign for smart meters carried out by Smart Energy GB focused on money people would save, thanks to direct feedback. By the end of 2019, Smart Energy GB will have spent £200m of consumer money to advertise smart meters back to them. This is equivalent to £6 per household. A colleague of mine once wondered how much more or less effective it would have been to just pay every household £6 to install a smart meter.

Privacy relates to the issues of how the information from smart meters can be stolen for fraudulent purposes, but also shared for commercial purposes. Smart meters which rely on mobile phone signal to transmit energy usage data raise concerns that households' data are not securely enough protected. In the United Kingdom the government commissioned the Data and Communications Company to design a smart meter communications infrastructure. This means that electricity usage data for every household will be available to third parties, as long as the customers' permission has been granted. Data privacy in Germany was at the centre of a set of protectionary guidelines as part of the Digitisation of the Energy Turna-round Act. This establishes transparent monitoring, the right to object to or correct user data and autonomous tariff selection. The Act also determined that meter reading intervals should be limited to protect consumers. The data transmission is protected through encrypting and limiting the extent of data transfer to minimal parties, while also prohibiting the use of this data for commercial purposes without explicit consumer consent.

With regard to safety, there have been consumer campaigns accusing smart meters of being dangerous because of the radio waves which are sent during remote readings. The 'no smart meter associations' have the highest level of participation in France (over four million people for 'stop Linky'), followed by 'Stop smart meters UK' and 'Stop contadores digitales' in Spain with about one million people and 'Stop smart meter Osterreich' in Austria and 'Stop contatore Enel' in Italy with a smaller number of participants. The main view of these associations is that smart meters contribute, along with other WiFi technologies, to radio exposure levels whose thresholds are set as too low by law. Governments of various countries have established that smart meters are not associated with health risks. For instance, according to the Public Health England, smart meters do not represent a threat to human health as the level of radio waves they produce is typically one million times less than the internationally agreed guidelines.

A book on the economics of smart meters

This book makes an attempt to frame smart meters in relation to their economic dimension. In presenting research on the costs and benefits of smart meters, I am aware that I am following the path left by other economists before me who endeavoured to objectivise a phenomenon at the expense of objectifying it. Smart meters are more than their costs and benefits as they have been charged with the potential to influence behaviour, but also could be a tool to measure how cyclical and repetitive human lives are. My escape from a bare cost-benefit approach in this book comes from the engagement in social theories which relate to smart meters and energy demand in the form of behavioural sciences and social practice theories. The former – stemming from environmental psychology – is suited to view the smart meter as informing individuals about consumption, hence potentially influencing their behaviour. The latter – originating from a branch of sociology – offers powerful lenses through which one can understand the structure of everyday life and, consequently, the timing of energy demand.

Chapter 2 presents an overview of the economics of smart meters. Their basic function consists of measuring levels of consumption. For this reason, a brief history of electricity meters is presented in which the reader can learn about how units to measure electricity demand changed over the years and the role of cost rationales in determining the various typologies of meters we have in our homes. This leads to a classification of main cost categories for smart meters, followed by categories of benefits. Chapter 3 presents estimates of costs and benefits of smart meter roll-outs in a set of European countries, including Germany, the United Kingdom, the Netherlands, Ireland, Hungary and France. It is highlighted that the amount of energy savings plays a vital role in determining whether the Net Present Value in any CBA is positive or negative because of the scale of roll-outs, which involve millions of users. Smart meters have been shown to provide significant benefits to Distribution System Operators in European countries where there are high levels of theft and commercial losses. Deferring investments in additional generation, transmission and distribution associated with peak electricity demand, the reduction of technical losses and reduction of the time of outages can also contribute towards compensating for the costs of smart meters. However, the extent to which these benefits are computed in CBAs depends on a number of factors which are considered in Chapter 3.

Chapter 4 presents a single-country case study on the economics of smart meters in Romania. It presents data from pilot studies carried out in the years 2015 and 2016 by the Romanian Distribution System Operators. The Romanian case study offers the opportunity to discuss in depth issues of smart-metering cost data and particularly capping smart-metering market price and depreciating smart meter costs.

Chapter 5 takes a step back and poses questions about the use of CBA as applied to the economic appraisal not only of smart meters, but also of energy policies in general. It is worth reflecting on the role of CBA in decision-making in energy as this is the most diffused economic tool for appraising the societal advantages and disadvantages of different policies and regulations. The economic case for smart meters has been assessed making use of CBA, and this emphasises its key role for policies with large investments of public resources. When CBA is performed by public authorities, as is the case for national roll-outs of smart meters, a wide array of societal impacts are mobilised. Chapter 5 frames CBA in historical perspective along with other appraisal and evaluation tools which have been used to assess the impacts of energy policies and regulations. The types of data and models which typically contribute to CBAs and the organisations involved in preparing them shape the different figures and uncertainties associated with this tool.

Smart meters alone do not deliver reductions in energy demand, yet their presence is supposed to prompt behavioural changes, thanks to the real-time feedback they display. They provide information that is otherwise not available to end users and combine that with accurate billing. Chapter 6 presents a meta-analysis of smart-metering trials and conservation effects. The literature does not produce any conclusive position around the effectiveness of smart meters. The review of smart-metering trials reveals the importance of sample sizes of the trials and sample selection methods in explaining variations in terms of net conservation effects. It is derived that over time there has been a steady decrease in net conservations effect coupled with an increase in the reliability of studies on smart-metering effects. This means that the economic case for smart meters based on net conservation effects is progressively losing ground.

Chapter 7 moves beyond traditional approaches of economics and system needs by introducing theoretical views on smart metering and energy demand from different strands of social sciences. The questions that Chapter 7 seeks to address relate to how energy demand can be analysed in terms of its structure and which conceptual positions can facilitate a better understanding of energy demand. Behavioural theories are used in connection with attitudes, preferences and choices around energy consumption, whereas social practice theories are better suited at explaining timing, space, rhythms and societal change in energy demand. The chapter links smart meters with research on people's activities and the timing of energy demand, key concepts of behavioural theories and social practice theories.

Chapter 8 concludes with some reflections on future developments of smart meters and changes in electricity generation and demand. The end of coal generation, along with the increasing cost-effectiveness of solar PV and battery storage, has led to the decarbonisation of the electricity grid in several developed countries. Many sources point to smart meters as an important piece of the puzzle in the digitalisation and decentralisation of the energy transition. Smart meters are seen as potential enablers for the challenge of demand-side flexibility when combined with demand-side response, storage battery and time of use tariffs. The chapter ends with six key messages provided by this book.

Note

1 Interpretative note on Directive 2009-72/EC concerning common rules for the internal market in electricity and Directive 2009/73/EC concerning common rules for the internal market in natural gas Retail Markets European Commission, Brussels (2010).
    http://ec.europa.eu/energy/gas_electricity/interpretative_notes/doc/implementation_notes/2010_01_21_retail_markets.pdf.

References

Alberini, A., Prettico, G., Shen, C., & Torriti, J. (2019). Hot weather and residential hourly electricity demand in Italy. *Energy, 177*, 44–56.

Allen, S. R., Hammond, G. P., & McManus, M. C. (2008). Prospects for and barriers to domestic micro-generation: A United Kingdom perspective. *Applied Energy, 85*(6), 528–544.

Barton, J., Huang, S., Infield, D., Leach, M., Ogunkunle, D., Torriti, J., & Thomson, M. (2013). The evolution of electricity demand and the role for demand side participation, in buildings and transport. *Energy Policy, 52*, 85–102.

Chmutina, K., & Goodier, C. I. (2014). Alternative future energy pathways: Assessment of the potential of innovative decentralised energy systems in the UK. *Energy Policy, 66*, 62–72.

European Commission. (2014). Cost-benefit analyses & state of play of smart metering deployment in the EU-27 (Report Number: 52014SC0189).

Grünewald, P., & Torriti, J. (2013). Demand response from the non-domestic sector: Early UK experiences and future opportunities. *Energy Policy, 61*, 423–429.

Hanna, R., Leach, M., & Torriti, J. (2018). Microgeneration: The installer perspective. *Renewable Energy, 116*, 458–469.

Hoenkamp, R., Huitema, G. B., & de Moor-van Vugt, A. J. (2011). The neglected consumer: The case of the smart meter rollout in the Netherlands. *Renewable Energy Law & Policy Review, 2*, 269–282.

Lo Schiavo, L., Delfanti, M., Fumagalli, E., & Olivieri, V. (2013). Changing the regulation for regulating the change: Innovation-driven regulatory developments for smart grids, smart metering and e-mobility in Italy. *Energy Policy, 57*, 506–517.

Meeus, L., & Saguan, M. (2011). Innovating grid regulation to regulate grid innovation: From the Orkney Isles to Kriegers Flak via Italy. *Renewable Energy, 36*(6), 1761–1765.

Torriti, J. (2012). Price-based demand side management: Assessing the impacts of time-of-use tariffs on residential electricity demand and peak shifting in Northern Italy. *Energy, 44*(1), 576–583.

Torriti, J. (2015). *Peak energy demand and demand side response*. Routledge Explorations in Environmental Studies (p. 172). Abingdon: Routledge.

Torriti, J., Hassan, M., & Leach, M. (2010). Demand response experience in Europe: Policies, programs and implementation. *Energy, 35*(4), 1575–1583.

Van Gerwen, R., Jaarsma, S., & Wilhite, R. (2006). Smart metering. https://www.leonardo-energy.org/

Van Gerwen, R., Koenis, F., Schrijner, M., & Widdershoven, G. (2010). Smart Meters in the Netherlands. Revised financial analysis and policy advice (No. KEMA--30920580-CONSULTING-10–1193). KEMA.

2   The economics of smart meters

Economics of electricity meters through history

Following discoveries in electromagnetism by the French André-Marie Ampère in the first part of the 19th century, the electromagnetic meter became an example of practical applications, along with the lamp, the dynamo, the motor and the transformer. The Hungarian Ottó Titusz Bláthy is considered the inventor of the induction electricity meter. The invention had market penetration in places where first mass applications of electricity occurred. For instance, when towns started replacing gas with electricity for the provision of public lighting, the need to measure volumes of electricity being sold ensued. Which volumes were measured and which units were billed at the time? In the United States, a meter patented in 1872 by Samuel Gardiner measured the lamps connected by a single switch and the time during which electricity was supplied to the lamps.[1] The companies providing electrical energy for lighting were also those manufacturing the meters, as in the last quarter of the 19th century there was no market for making meters.

Thomas Edison patented his electric meter in 1881. This happened following the introduction of the first electrical distribution systems for lighting using direct current, which introduced the possibility of selling electricity like gas. Inside the meter there was a copper strip which was weighed at the beginning and at the end of the billing period. The difference in weight of the copper strip represented the amount of electricity that had passed through. The meter was calibrated so that the bills could be measured in cubic feet of gas. Until the end of the 19th century, these were the most common electricity meters, despite the fact that they were difficult to read for the utility. Consumers were not even allowed to read them until Edison added a counting mechanism to aid meter reading.

The cost of the meter was a very significant criterion for exploring various versions of early days electricity meters. For instance, there was a meter with two pendulums which had two clocks and measured ampere-hours or watthours. The pendulum meter was considered too expensive and was replaced by motor meters in which the rotor speed would vary depending on voltage of electricity consumption.

Reducing costs was again a driver of innovation. The cost of the motor was considered too high and the induction meter replaced the motor. This was based on the findings of an Italian engineer, Galileo Ferraris. The first meters were mounted on a wooden base, running at 240 revolutions per minute, and weighed 23 kg. By 1914, the weight was reduced to 2.6 kg. Induction meters, also known as Ferraris meters, were manufactured in massive quantities across the world throughout the 20th century, thanks to their low price and excellent reliability.[2]

Despite the fact that for a long period the Ferraris meters were considered the cheapest technological metering option yielding adequate benefits in terms of reliability and functionalities, there have been variations with regard to the functions of the meters in different places. For instance, meters in North American and UK residential households are not capped in terms of capacity limits. This means that people can consume almost limitless amounts of electricity without experiencing any power cut. This is because the amperage provision in these countries is extremely generous, with 10 ampere meters allowing for much higher power than any household will ever reach at peak consumption. For example, Moore (1935) in 1915 visited a small village in the United States where meters were being installed inside cottages. He asked the engineer in charge of the installation: "Why did you install 10-ampere meters in these cottages when the maximum load is not more than about half an ampere?" The engineer replied: "Well, I found that I could get a 10-ampere meter for the same price as a 2-ampere meter". In other countries, for instance, in Southern Europe, meters amperage is much more limited or suited to the maximum power demand of households. In Italy, the capacity limit (*potenza impegnata*) is the power level indicated in the contracts and made available by the retailer. It is defined according to the customer's needs at the moment of the conclusion of the contract, depending on the type (and number) of electrical appliances normally used and, for domestic customers, also using the available information on the maximum demand level taken in each month. The total number of residential users with capacity limits in Italy is 28 million, and 90% of residential customers have a capacity limit of 3 kW. In addition, capacity limits are also applied to non-residential users up to 15 kW. In Spain, consumers must choose between one of the capacity limits ('*potencia contratada*') according to their needs, taking into account that if they request a higher capacity, the electricity bill will be higher and if they request less than they need, the meter will switch electricity off, as explained in Table 2.1.[3]

In Spain the fuse box acts as controller (*interruptor de control de potencia*) as it measures the level of power being used and cuts temporarily supply when the capacity limit is reached (Santiago et al., 2014). The fuse is in place in any of the smart meters installed in Spain (López-Rodríguez et al., 2013) (Table 2.2).

In Portugal, the approach to capacity limit is very similar to Spain and Italy (Guerreiro & Ferreira, 2015). In Romania, there are no (widely) used

Table 2.1 Capacity limits of electricity meters in Spain

Capacity limit (kW)	Characteristics
2.3	Studio flat without air conditioning
3.45	Small flat without air conditioning with small appliances
4.6	Medium size flat with air conditioning in some rooms + small appliances
5.75	Medium size flat with air conditioning + medium size appliances (oven, dryer)
> 6	Medium and large flats with air conditioning and large appliances

Table 2.2 Fuse intensity and maximum power in Spanish meters

ICP fuse intensity (Amps)	Monophase (kW)	Three-phase (kW)
5.0	1.5	3.464
7.5	1.725	5.196
10	2.3	6.928
15	3.45	10.932
20	4.6	13.856
25	5.75	17.321
30	6.9	20.785
35	8.05	24.249
40	9.2	27.713
45	10.35	31.177
50	11.5	34.641
63	14.49	43.648

meters with internal disconnection capabilities. However, each meter is under a connection contract which stipulates clearly the maximum power allowed. Each customer has a connection contract with a capacity limit. There are also fuses, usually electronic ones, which limit the absorbed current, but these devices are under the consumer control.[4]

With changes in electricity markets, the need to have multi-tariff meters with local or remotely controlled switches became higher. As a result, the maximum demand meter, the prepayment meter and the maxigraph were invented and introduced to the market by the turn of the century. The following development consists of meters which are connected with in-home display, can be read remotely and host two-way communication systems between the end user and the utility. At the time of writing this book, the metering technologies in place in most European countries still heavily rely on electromechanical meters, i.e. Ferraris meters. Electromechanical metering technologies require suppliers' intervention in order to read electricity consumption and are subject to a vast range of errors. Such outdated metering technology prevents the main functions of demand-side participation, including hourly metering reading, information feedback to customers via in-house displays, automated direct load control and two-way communications. While it is commonly agreed that a wider use of electric meters than what is currently experienced in Europe would create the bases for most of these functions, changes in government backing of smart meters (see Chapter 3); increasing costs associated with communications and security; and the lack of a liberalised metering market in most European countries, apart from Spain, Italy and the Netherlands can help understand why this replacement has not taken place in Europe to date.

Costs of smart meters

Assessing the costs of metering infrastructure is a complex process which requires systematic analysis of impacts across the supply chain. The costs associated with investments in changes to electricity metering infrastructure relate to retail through to distribution, transmission, the wholesale electricity market and ultimately to the end user (BEIS, 2019).

The main costs can be divided into three categories: meters, meter installation and the communications and data systems (including capital, installation and management). Stranded costs can also be an important cost component when considering the business case for investment.

When it comes to formalising the economic assessment of smart meters into a cost–benefit analysis, the European Commission identified seven steps: reviewing and describing technologies, elements and goals; mapping assets into functionalities; mapping functionalities into benefits; establishing the baseline; monetising benefits and identify beneficiaries; identifying and quantifying costs; and comparing costs and benefits.

All European Union member states have to follow the recommendations of the European Commission, although their cost–benefit analyses all varied, as it is explained in Chapter 3. To facilitate the take-up of this new technology, the European Commission published a Recommendation (2012/148/EU) to prepare the roll-out of smart-metering systems. The Recommendation provides step-by-step guidelines for member states on how to conduct cost–benefit analysis. It also sets common minimum functionalities of smart-metering systems and addresses data protection and security issues.

The Recommendation puts forward a four-stage approach consisting of four main areas. First, the recommendation to tailor to local conditions implies that pilot programmes and 'real-life' experience should be used in assumptions. The two recommended scenarios consist of business as usual and 80% roll-out by 2020. Second, cost–benefit analysis is supposed to follow 'seven steps', indicating costs to be incurred by the consumer and to be compared with long-term benefits. Third, a sensitivity analysis should identify 'critical variables' and analyse magnitude of the cost–benefit analysis outcome for the positive roll-out conditions (i.e. when benefits exceed costs). Fourth, performance assessment, externalities and social impact should be used to assess externalities (e.g. in terms of the environment and carbon implications) and the impact of public policy measures and social benefits (i.e. ensuring appropriate weighting factors).

In the implementation of this methodological approach, various factors need to be considered, including the distinction between economic and private costs and benefits, the period of analysis, the time profile of costs and benefits and the discount rate applied.

With regard to the distinction between economic and private costs and benefits, the principal purpose of the cost–benefit analysis is to determine whether a widespread roll-out of smart meters is economic for the country as a whole. In practice, certain participants may bear a disproportionate share of the costs, which by itself does not provide grounds for overturning the results. However, distribution issues are critical at the implementation stage, and in particular in the development of policy recommendations.

The period of analysis, including replacement of "new" assets (smart meters and communications equipment), is also a key consideration, and in any case, as is the terminal value of these assets at the end of the modelling period, where a residual value of assets will be present.

The time profile of costs and benefits, in particular in the period of the roll-out, assumes a very relevant role taking into account the life expectancy of the meter as well as the time scales of the roll-out.

With regard to the discount rate to be applied, the European Commission guidelines suggest that in most cases national cost–benefit analyses may not move away from standard social discount rates, ranging from 3.5% to 5.5% in different European member states.

Table 2.3 lists the categories of costs as identified in the EU 2012/148/EU Recommendation. As explained in Chapter 3, critical to the variation in units of costs are the choice of metering equipment, communications and IT technology, and local labour costs. In addition, revenue reduction, change in fossil fuel costs and emissions costs could also be accounted as benefits, depending on whether they generate net positive or negative impacts.

The costs of electronic meters have for long been a practical constraint to demand participation. Whilst production costs of some of the most inexpensive technological solutions have been decreasing over the last few years, rendering them affordable, three problems persist with regard to the costs of enabling the most expensive and advanced smart-metering technologies. First, the roll-out costs associated with start-up and implementation are significantly high. An ERGEG (2007) survey shows that in 11 European countries the penetration level was still below 7%. This made the marginal costs of roll-out very high compared to places like Sweden, Italy, Finland and France, where the penetration level is 20% or above. Second, regarding the benefits side, most of the debate around the economic potential of demand-side participation has been focusing on paybacks to consumers. On this point, the existing studies fall short of estimating the potentially wider

Table 2.3 List of costs (from EU 2012/148/EU Recommendation)

Type of cost	Category
Investment in the smart-metering system	CAPEX
Investment in IT	CAPEX
Investment in communications	CAPEX
Investment in in-home displays (if applicable)	CAPEX
Generation	CAPEX
Transmission	CAPEX
Distribution	CAPEX
IT maintenance costs	OPEX
Network management and front-end costs	OPEX
Communication/data transfer costs (e.g. GPRS)	OPEX
Scenario management costs	OPEX
Replacement/failure of smart-metering systems (incremental)	OPEX
Revenue reductions (e.g. through more efficient consumption)	OPEX
Generation	OPEX
Transmission	OPEX
Distribution	OPEX
Meter reading	OPEX
Call centre/customer care	OPEX
Training costs (e.g. customer care personnel and installation personnel)	OPEX
Restoration costs	Reliability
Emission costs (e.g. CO_2)	Environmental
Costs of fossil fuels consumed to generate power	Energy security
Costs of fossil fuels for transportation and operation	Energy security
Sunk costs of previously installed (traditional) meters	Other

impacts of innovation (Bilton et al., 2008). Most studies rely upon the evidence provided by pilot roll-out cases showing that – when fully informed – consumers tend to respond to higher energy prices. However, the long-term economic benefits and carbon emission reductions have not been fully unexplored. Third, the most advanced technologies are not cost-effective yet. For instance, advanced energy management systems in commercial buildings and process control systems in industrial facilities that can reduce load when needed are above reach for most commercial and industrial users, mainly because customer awareness of these tools is low. In turn, the price of these technologies is high due to the low level of market penetration, since the marketing infrastructure is in very early stages. In other words, the value chain from the device producer to the retailer is disadvantaged by low levels of cumulative installations and limited experience. This means that some of the costs estimated in economic appraisals may change. The example of Romania in Chapter 4 shows that the cost of a smart meter unit decreased over time. Conversely, in the United Kingdom, the Competition and Markets Authority (2016) found that installation costs are up to 50% higher than initially estimated because large suppliers incur higher costs than they would in a perfectly competitive market and suppliers in general do not have incentives to keep installation costs down.

Benefits of smart meters

Smart meters have been linked with several benefits, including lower metering cost, energy savings for residential customers, more reliability of supply, variable pricing schemes to attract new customers and easier detection of fraud. In addition, other benefits are foreseen in relation to distributed generation as the smart meter can be used to separately measure electricity delivered by distributed generation to the grid and the smart-metering communication infrastructure can be used to remotely control distributed generation. Demand-side response and dynamic pricing can be enabled via the smart meter. Reductions in meter reading operations and reductions in technical losses of electricity through higher efficiency in network operation can be considered as other main areas of benefits.

Table 2.4 lists the categories of benefits as identified in the EU 2012/148/EU Recommendation. Categories like consumption reduction are extremely relevant for cost–benefit analysis purposes because the small benefits are multiplied by millions of end users. In Chapter 3, smart meters have been shown to provide significant benefits to the Distribution System Operator (DSO) in countries with high levels of theft and commercial losses. Similarly, the scope to defer peak investment, reduce technical losses and reduce the time of outages will depend on the starting position of the country.

Table 2.4 List of benefits (from EU 2012/148/EU Recommendation)

Benefits	Sub-benefits
Reduction in meter reading and operation costs	Reduced meter operation costs
	Reduced meter reading costs
	Reduced billing costs
	Reduced call centre/customer costs
Reduction in operational and maintenance costs	Reduced maintenance costs of assets
	Reduced costs of equipment breakdowns
Deferred/avoided distribution and maintenance costs	Deferred distribution capacity investments due to asset remuneration
	Deferred distribution capacity investments due to asset amortisation
Deferred/avoided transmission capacity investments	Deferred transmission capacity investments due to asset remuneration
	Deferred transmission capacity investments due to asset amortisation
Deferred/avoided generation capacity investments	Deferred generation investments for peak-load plants
	Deferred generation investments for spinning reserves
Reduction of technical losses of electricity	Reduced technical losses of electricity
Electricity cost savings	Consumption reduction
	Peak load transfer
Reduction of commercial losses	Reduced electricity theft

A general question in considering the benefits of smart metering is presented here: is there a business case for investing in smart metering for all consumers? Many countries already employ some form of advanced metering for industrial customers, but the policy question here relates more specifically to smaller users, i.e. small and medium businesses and households. In a number of international studies, the costs of implementing smart metering in these sectors still outweigh the benefits when looking at the business case from either a supplier or network operator perspective (Capgemini, 2007; Carbon Trust, 2007; Ofgem, 2006). Where operational and demand response benefits are more integrated, the business case has tended to be positive as in the case of the California utilities; this requires separating financing from the customer via regulated charges. In Great Britain and Australia, alternative market models have been quantitatively explored in order to assess how changing responsibilities affects total costs and the cost/benefit ratio, as well as the allocation of costs and benefits across market actors. Findings suggest that there may be significant cost savings from a more coordinated regional rollout strategy. A return to a distributor-led metering system in Great Britain, however, is not the only way of achieving this; the regional franchise model could be a viable alternative. Competition in the tender process has the potential to drive innovation and the delivery of the least cost solution. The realisation of benefits can be negatively affected by at least two factors: suppliers' operational cost reduction not materialising and consumers' lack of uptake. With regard to suppliers' cost reduction, a National Audit (2018) report highlights that according to retailers the roll-out of the first generation of smart meters helped them reduce the number of manual meter readings and inbound contacts to call centres have been lower. According to the same source, it is less certain the extent to which suppliers will be able to reduce the cost of customer support to the extent set out in a BEIS (2016) cost–benefit analysis because of uncertainties around the functioning of the Data and Communications Company (DCC) system and the fact that suppliers will need to support a mixture of different generations of smart meters as well as traditional meters and not just last generation meters as envisaged in BEIS (2016), hence increasing capital and operational costs. The most recent UK cost–benefit analysis (BEIS, 2019) includes customer benefits (including energy savings), supplier benefits, demand-shifting benefits, network benefits, environmental benefits and unquantified benefits.

Other risks, benefits and costs

The standard classification of costs and benefits provided in Sections 2 and 3 does not include several categories of risks, costs and benefits, which complicates the picture when it comes to understanding the real implications of smart-metering implementation.

To begin with, smart meters are supposed to cover several transdisciplinary areas, including security of supply, accurate energy bills, promoting awareness of energy consumption, leading to increased awareness of the cost of energy use, promoting energy efficiency behaviours, simplifying the transition from prepayment to credit, remote meter reading, encouraging use of demand-side management, better data for demand/load forecasting and facilitating wider use of microgeneration (Watson, 2009). The examples of cost–benefit analyses in Chapters 3 and 4 capture elements of security of supply with the quantification of technical losses, and remote meter reading through reduction in meter reading and operation. However, there are risks, costs and benefits associated with the other impacts which are typically not assessed.

This section sets out to capture additional elements of risk, costs and benefits which are not included in typical cost–benefit analysis practice on smart meters. It does so by including a discussion based on the UK roll-out experience and prospective discussion around appraising interoperability.

UK experience

In the 1980s the South Eastern Electricity Board (SEEBOARD) had designed the first smart meter in the world which was called the Credit and Load Management Unit, which was (successfully) trialled by SEEBOARD in 300 homes commencing in 1983 for approximately two years. Attempts to commercialise these early days smart meters were not successful at a time when the focus in the electric industry was on privatisation.

In the mid-2000s, smart-metering plans started taking place in the United Kingdom. Around the same period in Italy in 2000 Enel designed and tested a national programme for smart meters. It ran a pilot study and then by 2008 had installed 32 million smart meters (Alberini et al., 2019). Meanwhile the Directive 2006/32/EC on Energy End-use Efficiency and Energy Services did not mandate smart meters, but pointed the way and provided a lobbying base for meter manufacturers and meter services providers. Later Directive 2009/72/EC, which was part of the EU Third Package, went much further by requiring member states to implement "intelligent metering systems that shall assist the active participation of consumers in the electricity supply market", provided that smart meters are "subject to an economic assessment of all the long-term costs and benefits to the market and the individual consumer and of which form of intelligent metering is economically reasonable and cost-effective and which timeframe is feasible for their distribution" (Torriti, 2010).

In 2005, Energywatch (2005), which at the time was the statutory consumer body, initiated a debate by publishing a report which reflected concerns about inaccurate estimated bills and suggested that smart meters of the type being installed in Italy at that time could offer a technology solution to this problem.

Smart meters entered the main policy agenda as part of the government's commitment to greening the electric industry. In 2007, a White Paper envisaged that "within the next 10 years, all domestic energy customers will have smart meters with visual displays of real-time information..." (DTI, 2007).

In parallel the government invested £9.75 million for the Energy Demand Research Project (EDRP), a set of trials which involved 18,370 households with smart meters. This process resulted in commissioning four essentially independent trials being conducted and analysed by energy suppliers (EDF, E-ON, Scottish Power and SSE) and their academic advisors. The EDRP showed that smart meters *per se* do not provide energy savings and net conservation effects of 3% only take place when the smart meter is combined with an in-home display. The EDRP trials took place in the context of a changing policy environment. They were carried out in parallel with two public consultations on the functionality and implementation of smart metering. At the start of the trials, the government had made no commitment to national adoption of any form of smart metering, but the passage of the Climate Change Act in 2008 paved the way for it.

The government took the decision to mandate a roll-out of meters in 2009. The new coalition government confirmed the decision to roll out smart metering when they took office in 2010 – that is a year and a half before the results of EDRP were published in 2011. The British roll-out of smart meters acquired unique characteristics due to the centralised communications system and because it is led by suppliers (as opposed to DSOs). To make smart meters interoperable between energy suppliers, the government set out to connect them to a central data and communications infrastructure, i.e. the DCC. The main alternative would have been, as it happens in most European countries (see Chapter 3), to make use of a power line carrier (PLC) using the distribution network as a data communications medium. According to BEIS (2016), one of the main benefits of the DCC system is that in its absence, smart meters would only realise those switching benefits that the analysis has identified to be realisable in the pre-DCC context for the domestic sector (£0.8 per smart meter per year).

The choice of a supplier-led smart-metering roll-out was a consequence of the "supplier hub" model conceived by Ofgem in the second part of the 1990s as part of fostering mass market retail competition. This means that suppliers are responsible for metering. In 2009, the Department considered adopting a Distribution Network Operator (DNO)-led delivery model for smart meters, but decided against this because it would take additional time to re-regulate the metering market, following the decision of giving energy suppliers responsibility for meters in 2003–2004 (National Audit Office, 2018). This was already a significant difference from other countries, like Italy, for instance, in which the DSO owned the meters. In principle, having several suppliers with an interest to keep the costs of metering down and yet offering higher quality of metering services to customers (e.g. via differentiated tariffs) would, in turn, create the market conditions for a competitive metering sector. However, as this book demonstrates, metering is not a commodity service, but a fundamental infrastructure asset. The UK government also performed some analysis on a DNO-led roll-out (House of Commons Energy and Climate Change Committee, 2015). However, the analysis relied on the same cost of capital for smart meters (10%) used for the supplier roll-out and assumed that even a DNO-led roll-out would make use of a central communications system. As a result, the option of a DNO-led roll-out was discarded on the basis that there would be no significant benefits.

The first cost–benefit analysis in 2007 was contracted by the British government to Mott MacDonald (2007). It estimated a negative Net Present Value in the region of £4 billion, as legacy meters and developing the communications infrastructure were considered extremely costly. Successive analyses by the Department of Energy and Climate Change (DECC, 2014) subverted this figure and pointed to a net benefit of £ 6 billion (see Chapter 3). As mentioned above, in 2007 the British government envisaged that the national roll-out of smart meters would be completed in ten years.

Some critics of the UK government smart-metering plan suggested that the roll-out should be run by DNOs instead to a basic specification for new installations and replacement meters, with DNOs responsible for backhauling the data (Henney, 2013). According to this view, the costs of the basic smart meters and communications would be socialised, while DNOs and other companies would give customers the option of installing higher functionality meters and in-home displays and pay the incremental costs.

Distributional effects are also difficult to address in a cost–benefit analysis framework (see Chapter 5). In national campaigns accompanying the smart-metering roll-out, it is often mentioned that benefits are spread across all end users, with specific reference to the volume of energy savings being comparable to providing energy for entire cities in the United Kingdom. However, the same costs of smart meters and the benefits associated not only with conservation effects but also in terms of demand management are not equally distributed. For instance, the Carbon Trust (2007) ran a field trial of advanced metering for small and medium enterprises. Overall there was a large benefit for larger companies, but a disbenefit for smaller companies of £110/meter/year from which one might infer there was likely to be a disbenefit for domestic customers. Intuitively, smart meters will have different impacts on different groups, but those differences have not been quantified. There are significant challenges that prevent it from measuring the level of energy savings for specific groups of consumers.

In 2006, Ofgem (2006) published a report on domestic metering innovation stating that a fundamental economic barrier to a market roll-out of smart metering, consists of the fact that the benefits arising from it are split between the suppliers, the DNOs and the customers. Chapter 4 will look further into distributional effects through a case study.

In 2013, it was forecasted that it would cost £11.0 billion and provide benefits of £17.7 billion over the period 2013–2030 (DECC, 2013).

The 2016 analysis reduced the estimated benefits of the programme to £16.7 billion, giving a net economic benefit of £5.7 billion (BEIS, 2016). Chapter 3 compares the different UK cost–benefit analyses and provides in contrast with other European analyses of smart-metering roll-outs. A National Audit Office (2018) report identifies several areas in which the UK government's cost–benefit analysis is likely to have understated the costs of the smart meter rollout. The report estimates that costs have increased by at least £0.5 billion since the BEIS (2016) cost–benefit analysis. The £0.5 billion consists of £0.3 billion in relation to the increasing costs of the DCC and £0.2 billion in relation to the cost of the technical solutions required to enable the full functionality of the system in the final 3.5%–5% of properties.

Other European countries' experience

In 2012, the European Commission recommended that member states should embark in smart-metering roll-out plans which should achieve 80% penetration by 2020. Following the ambitious targets of the Commission's Recommendation on preparations for smart-metering roll-out (2012/148/ EU), in 2014, the EU-wide deployment target was decreased to 72%. This was necessary because only about half of the member states committed to meeting the 2020 deadline. What is the actual state of implementation in European countries? The United Kingdom, France, the Netherlands and Sweden are in line with the EU targets, but not Germany, as it is explained below.

In Austria the rollout of smart meters can be expected to commence in 2016 and 2017 and will have at least 95% of all metering points by the end of 2019. The energy regulator estimates that about 200,000 end users are provided with smart meters. There are plans for a complete rollout of 2.5 million smart meters. In Denmark it was estimated that in 2011 around 50% households had remote reading (Torriti & Grunewald, 2014). In Finland more than two thirds of targeted smart meters have been installed. In France, the rollout of smart meters started in 2013. In Germany smart meter penetration is estimated to be around 1 million. The delay of the smart-metering roll-out compared with other major European countries is down to the fact that a cost–benefit analysis carried out by the German government in 2013 concluded that the rollout by 2020 in accordance with the rollout scenario recommended by the EU would not be beneficial in terms of key macroeconomic indicators. In 2014 the French government announced that the full nationwide rollout of 35 million smart meters which was previously (in 2011) set to 95% by 2020, with an investment of €5 billion, had to be adjusted to 90%. Italy completed the roll-out of smart meters in 2011. The rollout of smart meters in the Netherlands was planned in two stages: a small-scale rollout between 2012 and 2014 and, after 2014, the large rollout ensues with a target of 80% by 2020.

Interoperability

One issue that emerges as extremely relevant to the economics of smart meters consists of their interoperability. Interoperability can be defined as the ability of a system to exchange data with other systems of different types. In other EU countries, analyses on the implementation of a smart meter roll-out have often been coupled with or followed by analyses on the impacts of different Information Communication Technologies (ICT).

This section describes the problem, potential options and issues around comparison of options and methods associated with different interoperability options.

Most of the smart-metering roll-out in Europe do not envisage how core smart-metering functionalities will be achieved and do not set targets for interoperability. However, different interoperability levels associated with smart-metering technology will trigger different impacts in terms of costs and benefits for a variety of stakeholders, including DSOs, electricity retailers, smart-metering manufacturers, ICT companies and end users. The interoperability of systems and devices will be a driving factor in determining the economics of any smart-metering roll-out programme.

From a regulator's perspective, the national roll-out of smart meters lends itself to questions of how interoperability will be achieved. With regard to interoperability, the EC Recommendation 2012/148/EU defines three core smart-metering functionalities: providing readings directly to the customer and any third party designated by the consumer, updating the readings frequently enough to allow the information to be used to achieve energy savings and supporting advanced tariff systems. Member states need to consider which interfaces enable to achieve these interoperability functionalities, as suits the relevant market arrangements and associated costs and benefits. In principle the functionalities might be achieved through interface standards defined by regulation, but also without the need for regulatory intervention (e.g. through self-regulation).

The economic problem facing any regulator is how to achieve functionalities for interoperability in an effective and sustainable manner for the electricity market.

In order to ensure sustainable levels of interoperability, regulators typically need to identify available options for interoperability; review technologies and applications associated with standards for interoperability; assess the impacts of different options; consider market and non-market stakeholder opinions on different standards; and decide whether to intervene (or not) through regulation, self-regulation or other means.

Separating analyses on different options for interoperability from analyses on national rollouts is what several governments in Europe did in order to assess the costs, benefits and risks associated with each interoperability option.

The EU Third Package underlines the importance of ensuring the interoperability of the intelligent metering systems by defining functionalities.

However, it is down to member states to define which interfaces should support the delivery of these functionalities. The European Commission suggests that, for instance, in order to achieve interoperability, interfaces between components of the smart meters might have to be based on open standards (as defined in the European Interoperability Framework).

ICT interfaces for smart metering are novel and present challenges which were not entirely thought through before. A study performed by the European Smart Grids Task Force Expert Group 1 on Standards and Interoperability in 2015 found that various member states had not taken extra measures needed to reach full interoperability and revealed several potential areas of risk from the viewpoint of interoperability. This is because some current smart-metering standards were not available when a deployment was planned and additional specifications (use cases, data definitions, companion standards) are not defined in the majority of member states' rollouts.

Ambiguity around interoperability across the end-to-end smart-metering process may have a negative long-term impact on the economic outcome of different smart-metering programmes. This could lead to a lack of economies of scale and innovation in customer services as smart metering should act as an enabler of additional services. For example, Chapter 4 points to large differences (250%) between the investment unit costs per smart meter presented by DSOs in different projects. The lack of interoperability targets may lead to further discrepancies in market prices of ICT connected to smart-metering devices, hence creating distortions in the market, halting competition in the smart-metering market and stalling innovation in the area of ICT.

At the same time, very stringent high standards and applications may lead to unaffordable costs, whereas low levels of interoperability may potentially bring about high costs in the further future when it comes to transitioning to smart grids.

Interoperability options comprise regulatory options, self-regulation and 'doing nothing'. Should standards be defined through regulation, the options may be shaped by where interoperability points are set. Maintaining the flexibility to update meters with new functionality remotely will minimise disruption for DSOs. On the other hand, concentrating all intelligence and control in the smart meter will oblige DSOs to conduct more frequent visits from field engineers who need direct access to the meter point.

In practical terms, regulating interoperability will involve a set of sub-options consisting of systems with higher/lower interoperability. There are a number of options currently for the communications outside the home, i.e. between the metering gateway and the power distribution network, utility, operators and any other authorised parties. A regulated higher standard of interoperability may increase the ability of the system to work well with other systems. According to the Smart Grids Task Force Expert Group 1 on Standards and Interoperability, the steps for defining what the sub-options will be may consist of functional analysis and creation/selection of use cases, selection of standards and technical specification, profiling based on standards/specifications and interoperability testing. ERGEG (2010) notes that higher interoperability standards through regulation may decelerate the roll-out of smart meters.

The alternative to regulating standards could be self-regulation. Self-regulation may set interoperability standards that ease the continuity of present investments into a future smart-metering landscape. Setting inter-operability at a point that minimises the cost and complexity of compliance with the final framework may become the main criterion. Under self-regulation, it is likely that every DSO will aspire to see the knowledge and capital it invested in the pilots carried over into its national rollout plans. In the search for a compromise through self-regulation, however, DSOs will need to come together to recognise that how interoperability standards are defined should not just address short-term cost protection. They will also need to safeguard the industry against the longer-term costs of upgrading the smart meter network to incorporate smart grid capability.

Under the 'do nothing' option, smart meters are rolled out according to the three common minimum functionalities on interoperability for smart-metering systems recommended by the EC (Recommendation 2012/148/EU), and defined in principle for electricity metering. These are generally in line with available and coming standards, and target the empowerment of the final customer. However, the Smart Grid Task Force encouraged the use of additional or companion specifications which describe how standards or technical specifications are applied to support the requirements of national infrastructure.

Carrying out a full societal cost–benefit analysis of the options is challenging for three reasons. First, the self-regulation option may require strong assumptions around which standards will be agreed among DSOs and which volumes of smart-metering implementation will occur. Second, for options regulating standards, the costs may be highly uncertain especially in relation to untested application profiles. Third, the benefits of different application profiles will be difficult to quantify and monetise unless some key benefits (e.g. more flexibility) are operationalised and strong assumptions are made around how these benefits materialise in the future (e.g. value of peak load shift in a smart grid).

As an alternative to cost–benefit analysis, multicriteria analysis may be used in combination with cost–benefit analysis as a way to take into account key criteria. Indicatively, criteria related to interoperability could consist of flexibility, innovation, competition and cost-effectiveness.

Convincing customers

Smart meters may or may not achieve any type of behavioural change, as their sole presence does not guarantee that people are nudged towards changes in consumption (Shove, 2015). Non-effects or neutral impacts of smart meters are examples of smart-metering benefits which may not materialise. However, more damagingly, there have been concerns around disbenefits or negative impacts of smart meters. These relate to privacy and radiation and have been covered extensively by the media and some research work in the Netherlands (AlAbdulkarim et al., 2012). Customers' acceptance of smart metering is becoming critical to the UK roll-out which (unlike other European countries) is led by suppliers and not DSOs. Energy suppliers stated that they will only be able to install smart meters in around 70%–75% of homes and small businesses by 2020 because of limited consumer interest and delays to the second generation of smart meters. These estimates may even be too optimistic if suppliers encounter further difficulty persuading consumers to accept installations (National Audit Office, 2018). This is problematic because according to BEIS estimates, each year of delay in completing the rollout reduces net benefits by around £150 million. There have been several attempts to understand aversion to smart-metering roll-outs. For example, a Belgian study of consumer preferences with regard to impact on the comfort and privacy level, functionality, visibility, cost savings and investment outlay finds that preferences do not align with smart-metering characteristics (Pepermans, 2014). The same study suggests that without careful preparation, a mandatory or voluntary roll-out of smart meters risks being unsuccessful because device characteristics do not meet consumer needs.

In general, the academic literature associates this lack of acceptance with negative mediatic exposure of smart meters and with the 'stop smart meters' campaigns mentioned in Chapter 1. For instance, a study of newspaper reporting in seven US states and one Canadian province showed that patterns of opposition are higher where the roll-out of smart meters is rapid and without an opt-out provision. Equally, technological differences (e.g. wired versus wireless) may contribute to levels of public opposition, and challengers to incumbent parties of either the right or left may also contribute to public opposition (Hess, 2014).

Smart meters and flexibility

Smart meters enable the deployment of tariffs that vary by time of day (e.g. Time of Use tariffs). Time of use tariffs are designed to encourage consumers to exercise their flexibility and shift their consumption away from times when the demand on the network is high (i.e. peak periods) (Torriti, 2015).

The importance of flexibility in electricity demand is often connected to the integration of intermittent renewables in the supply mix as well as high penetration of electric vehicles and electric heat pumps. These will challenge the balancing of demand and supply. At the same time, there are opportunities to reduce the costs of electricity if smart systems and storage are utilised to flex demand at times when this is high. In a nutshell, demand-side flexibility is portrayed as a win-win solution as consumers will help balance the grid in return for lower bills if they take advantage of smart appliances and smart tariffs (Torriti, 2019).

Smart meters are seen as a necessary condition for the large penetration of time of day tariffs for residential users (Torriti, 2012). From an economic perspective, the effectiveness of the Time of use tariffs is constrained by energy users' ability and motivation to change consumption based on price signals. Social science researchers would argue that the engagement of consumers with dynamic tariffs depends on the way to which everyday life and household rhythms can be aligned with the new tariffs (see Chapter 7).

In the United Kingdom, peak load shifting accounts for around 5% of expected smart-metering benefits, based on an assumption that time of use tariffs would have already been in common use by 2018 (BEIS, 2016). To put this in perspective, the UK government's Clean Growth Strategy presents estimates on levels of electricity demand flexibility for the future. In the impact of the 2032 pathway calculations for an 80% renewables future, electricity demand is projected to increase by 3% (i.e. 10 additional TWh), with an increase in peak demand of 4% (i.e. 2.8 GW) compared with current levels. The extra flexible capacity is expected to originate from demand-side response (4.9 GW), storage (0.3 GW), clean generators (0.5 GW) and fossil fuels (1.2 GW). The increase in peak demand is argued to arise from the uptake of electric vehicles and heat pumps. Implicit demand-side response would consist of shifting the charging time of most electric vehicles to overnight combined with smart heat pumps. As a result, the potential benefits of peak load shifting in the 2020s may have increased since the BEIS (2016) cost–benefit analysis, because the number of electric vehicles expected to be in use has increased significantly since then.

In Chapter 8 different tariffs which are supposed to foster demand-side flexibility under future price arrangements are presented. In this section, smart meter functionalities are introduced, as they relate to potential flexibility options. Smart meters are primarily designed to record electricity consumption per 30-minute period, submit readings to suppliers and/ or DSOs and provide energy usage feedback to consumers. Additional functions relate to direct load control. Smart meters can provide a load limiting function through which a load limit value (i.e. a threshold value) is established. This enables each individual smart meter to switch off supply if the threshold is breached. The load limiting function is expected to operate as follows: if active power import (kW) exceeds a configurable threshold over a period of time, smart meters are expected to notify the supplier and the consumer (via the in-home display). If load limiting is enabled, then immediately after notifications, the smart meters are expected to interrupt supply to the property. Either consumers can reinstate supply using the in-home display or supply is automatically restored after a configurable period of time. Smart meters are also expected to record the maximum power imported for each 30-minute period. Potentially this information could be used by suppliers or DSOs to enable maximum-demand tariffs.

Smart meters versions include the requirement to support different types of pricing approaches and tariffs. In addition to time of use tariffs, which are explained in Chapter 8 (Section 3.1), block pricing and time of use with block pricing can also be enabled by smart meters. These pricing methods are applied to both credit and prepayment payment methods.

An additional feature consists of the ability to perform direct load control of up to five appliances. Control is expected to work based on an in-built calendar of schedules or by remote request (e.g. from supplier or network operator).

Smart meters with twin-element variants can measure and bill for two separate circuits: primary supply and secondary supply. Primary circuit consists of the main supply to the household and secondary, at a smaller rating, would be for a separate set of loads (e.g. hot tub, heating or EV charger).

Distributional effects

National rollouts of smart meters are designed to install these devices in every household irrespective of their income and status. On the one hand, this means that there are no distortions in terms of access to information, possibilities of switching suppliers and benefiting from flexible tariffs. On the other hand, the cost of smart metering is the same for all categories of residential consumers, and this has implications in terms of which proportions of electricity bills are paid for. In other words, the same smart-metering cost, which is typically spread across bills for several years, will be more difficult to afford for lower income consumers than others. For instance, in the UK smart-metering costs represent about 2% of energy bills. In essence, the unit cost of the smart meter is intimately related with trends in bill costs. Smart metering may positively affect the fuel poor primarily through provision of better feedback to assist with energy management, and through simplifying and encouraging prepayment and supplier switching.

> Some positive aspects of smart metering emerge in relation to fuel poverty. These are: the use of smart meters to facilitate prepayment (which in itself helps energy management), and to remove any reason for charging higher unit prices for prepayment customers; the deployment of energy displays and clear, accurate written information as an integral part of smart metering systems, to help customers visualize their energy use and alert them to unusual usage patterns; and the use of smart metering data in conjunction with advice. On a more speculative note, fuel poor households could benefit from preferential access to cheap electricity, for example at times of abundant wind supply, especially if they use storage heaters and have hot water tanks.
>
> (Darby, 2012, 105)

In 2015 more than 2.5 million households lived in fuel poverty in England – the equivalent of 10% of households. Fuel poor people are in significantly greater need now than ten years ago, as the amount by which a consumer's energy needs would need to be reduced in order for them not to be fuel poor in 2015 was 39% higher than in 2005 (in real terms). Over the same period, the average energy bill increased by 27%. This means that either people in fuel poverty were less able than the average consumer to reduce their energy needs or they consumed less than their energy needs. The aggregate fuel poverty gap for England is at £884 million. The main factors for vulnerable consumers are low incomes and energy (in)efficiency of the homes they live in (for instance, the proportion of households in fuel poverty has the lowest levels of insulated cavity walls). The most vulnerable consumers (so called 'fuel poor') are defined as having fuel costs that are above the national median level and being left with a residual income below the official poverty line after spending for fuel costs. Household income, household energy requirements and fuel prices are three important elements in determining whether a household is fuel poor. It has been noted that under different fuel poverty measurement methodologies, the number of households defined as fuel poor could be even double compared with current government estimates. The price difference between the average Standard Variable Tariff default deal from the six largest suppliers and the cheapest tariff in the market recently reached over £300 in the year 2017. Vulnerable consumers are more likely to find themselves on a Standard Variable Tariff deal and, because of their circumstances, can suffer more harm as a result. Capping prices is only one of the many ways of avoiding that vulnerable consumers pay 'too much' for their energy bills. Other examples consist of increasing incentives on energy efficiency, strengthening and supporting the Home Energy Conservation Act and enhancing the level of support to local councils' initiatives. Introducing new sources of power increases the need for flexible generation and consumption. Analysis by UKERC (2017) suggests that the cost of integrating these sources of power remains relatively modest (between £5 and £10 per MWh), but could increase substantially if smart meters fail to be adopted at significant scales.

Smart meters per se do not resolve energy vulnerability and fuel poverty issues, and it is extremely rare to see any evidence of distributional effects in economic appraisals. However, some vulnerable consumers may need further support if they are to realise the full benefits of smart metering. In the United Kingdom a government-funded study found that groups who are more likely to be in vulnerable circumstances were broadly consistent with those of non-vulnerable consumers and, in some cases, were better off in terms of cost savings (Ipsos MORI, 2018).

Smart meters in principle can be integrated with measures which are aimed at addressing issues of consumer vulnerability. Two examples consist of energy cooperatives and social tariffs.

Energy cooperatives in several developed countries have already adopted smart meters as a tool to inform, communicate and educate around energy demand (Akadiadis et al., 2017). Smart meters can supply information on household consumption, which was not previously available to energy cooperatives, hence facilitating data analysis, even though consumers typically do not sign contracts with energy cooperatives. The experience in Spain and Italy shows that some of the most useful advice provided by energy cooperative consists of whether Time of Use tariffs are convenient for that specific household. For example, In Italy, smart meters allowed policy-makers to reconsider the design of the measures in order to guarantee a "minimum level of consumption" (compatible with the welfare requirements) to all the households. In Germany, households not able to pay their bills had their power demand reduced to 1,000 W instead of being completely cut-off from electricity so that a basic energy amount could still be drawn (Pye et al., 2017). Smart meters offer additional functionalities which can prevent or reduce the risk of self-disconnection. Alerts can be sent in case of low credit or high consumption and a wider range of top-up channels are available, including mobile phone and online top-ups (Ofgem, 2019).

Progressive tariffs are often introduced with the purpose of reducing electricity consumption, load and independency, but also as a social instrument to redistribute rising costs of electricity from low consumption to high consumption households.

In principle EU legislation favours socially aimed tariffs. According to the EU Directive 2006/32/EC on energy end-use efficiency and energy services "Member States may permit components of schemes and tariff structures with a social aim, provided that any disruptive effects on the transmission and distribution system are kept to the minimum necessary and are not disproportionate to the social aim".

In practice, examples do not abound. Since 2007 Italian consumers with lower income and/or the need to use high power consuming life-saving health machineries receive a social discount (*bonus sociale*) for their energy consumption.

A similar scheme for vulnerable consumers exists in California, where life-line rates were introduced in 1975 due to social reasons. The California Public Utilities Commission made the new progressive rates for basic consumption compulsory (Dehmel, 2011). An algorithm was utilised to calculate the rate: 50%–60% of the average consumption in summer times and 60%–70% in winter times for households with electric heating (Hennessy & Keane, 1989).

The examples above show instances in which core power demand has been changed. The main indication from these international examples is that the regulatory activity associated with changes in power demand levels would be significantly increased. For instance, in Italy, as part of the 2017 reform, not only were rules around capacity limits modified, but also heat pump tariffs and welfare benefits for vulnerable consumers. In its impact assessment in 2016, the Italian regulator (ARERA, 2016) suggests that each year there should be a review monitoring developments associated with capacity limits. Whilst we did not find evidence of the costs for different market actors of future changes to core capacity, with regard to benefits to consumers, two are often cited: improving comfort and energy efficiency by installing new electric appliances; and bills savings based on fine-tuning the excessive levels of power made available in relation to end users to power which is actually used.

Risks

Risk aversion, risk of technological obsolescence and unawareness of price-based tariffs are some of the most cited behavioural risks associated with smart-metering usage.

Risk aversion can potentially affect customers' preferences when it comes to choosing between flat rates and dynamic pricing. When selecting the type of rate, customers may focus on the downside risk that their bills can rise if they go on the rate, than on the upside potential that they might save money either by changing their load shape or from high cost to low cost periods. Research by Schultz and Lineweber (2004) shows that customers who experience time varying rates have high levels of satisfaction. Regardless of customer satisfaction rationales, suppliers often fight against risk aversion by using default pricing, which leads to much higher customer enrolment, rather than opt-in enrolment.

The risk associated with technological obsolescence is shared by end users and policy-makers. Perceptions on the useful life of price sensitive control technologies drive the risk of technological obsolescence. People might think that smart meters could shortly be replaced with newer technologies (Torriti et al., 2011). The risk of technological obsolescence is also extended to the enabling technologies. Most of these concerns enhance doubts around the capacity to recover the costs of investments before they need to be substituted. On this issue, the role of government as a risk-taker is key. On the one hand, a strong presence of the state in the implementation of advanced metering technologies will favour the decrease of prices. On the other hand, such presence in the metering market may induce a rate-based deployment of metering technologies; hence, this poses barriers to increased private investments.

The risk that customers are either not fully aware or not capable of understanding the change involved in price-based mechanisms can strongly undermine the successful outcome of some of the foreseen benefits of smart meters. The historical lack of customer exposure to dynamic pricing represents a substantial obstacle to their participation. The risk of inertial behaviour can be justified by a long-standing predominance of flat rates imposed by the incumbent supplier and ultimately contributes to low participation rates in voluntary programs.

Notes

1 http://energyaction.ie/fuel-poverty-conference/history-of-the-electric-meter/
2 https://www.smart-energy.com/top-stories/the-history-of-the-electricity-meter/
3 https://www.mipodo.com/blog/eficiencia-energetica/potencias-electricas-normalizadas/
4 Information obtained from Romanian energy regulator (ANRE).

References

Akadiadis, C., Savvakis, N., Mamakos, M., Hoppe, T., Coenen, F., Chalkiadakis, G., & Tsoutsos, T. (2017). Analyzing statistically the energy consumption and production patterns of European REScoop members: Results from the H2020 Project REScoop Plus. [Accessed online on 1st March 2019]. Retrieved from https://www.narcis.nl/publication/RecordID/oai:tudelft.nl:uuid%3A5927c204-a005-4ec7-9b54-32a0d12eb2d7

AlAbdulkarim, L., Lukszo, Z., & Fens, T. W. (2012). Acceptance of privacy-sensitive technologies: Smart metering case in The Netherlands. In *Third international engineering systems symposium CESUN*. Delft University of Technology, The Netherlands

Alberini, A., Prettico, G., Shen, C., & Torriti, J. (2019). Hot weather and residential hourly electricity demand in Italy. *Energy, 177*, 44–56.

ARERA. (2016). Relazione AIR - Riforma delle tariffe di rete e delle componenti tariffarie a copertura degli oneri generali di sistema i clienti domestici di energia elettrica. [Accessed online on 1st March 2019]. Retrieved from https://www.arera.it/it/elettricita/auc.htm

BEIS. (2016). Smart meter roll-out cost benefit analysis. [Accessed online on 1st March 2019]. Retrieved from https://assets.publishing.service.gov.uk/government/uploads/system/uploads/attachment_data/file/567168/OFFSEN_2016_smart_meters_cost-benefit-update_Part_II_FINAL_VERSION.PDF

BEIS. (2019). Smart meter roll-out cost benefit analysis. [Accessed online on 1st March 2019]. Retrieved from https://assets.publishing.service.gov.uk/government/uploads/system/uploads/attachment_data/file/831716/smart-meter-roll-out-cost-benefit-analysis-2019.pdf

Bilton, M., Leach, M., Anderson, D., Tiravanti, G., Green, T. M., & Prodanovic, M. (2008). Response to Ofgem's consultation on domestic metering innovation. London: Imperial College London. [Accessed online on 1st March 2019]. Retrieved from http://www.ofgem.gov.uk/MARKETS/RETMKTS/METRNG/SMART/Documents1/13357-Imperial_Centre_response.pdf

Carbon Trust. (2007). Advanced metering for SMEs carbon and cost savings. [Accessed online on 1st March 2019]. Retrieved from https://www.carbontrust.com/media/77244/ctc713_advanced_metering_for_smes.pdf

Competition and Markets Authority. (2016). Energy market investigation: Final report.

Darby, S. J. (2012). Metering: EU policy and implications for fuel poor households. *Energy Policy, 49*, 98–106.

DECC. (2013). Smart meter roll-out for the domestic and small and medium non-domestic sectors (GB). [Accessed online on 1st March 2019]. Retrieved from https://assets.publishing.service.gov.uk/government/uploads/system/uploads/

attachment_data/file/276656/smart_meter_roll_out_for_the_domestic_and_small_and_medium_and_non_domestic_sectors.pdf

Dehmel Christian (2011). Progressive electricity tariffs in Italy and California – Prospects and limitations on electricity savings of domestic customers, *ECEEE Summer Studies*. [Accessed online on 1st March 2019]. Retrieved from http://proceedings. eceee.org/papers/proceedings2011/2–275_Dehmel.

Department of Trade and Industry. (2007). Meeting the energy challenge. [Accessed online on 1st March 2019]. Retrieved from https://www.gov.uk/government/publications/meeting-the-energy-challenge-a-white-paper-on-energy

ERGEG. (2007). Smart metering with a focus on electricity regulation, Brussels: European Regulators' Group for Electricity and Gas, Ref: E07-RMF-04-03.

ERGEG. (2010). An ERGEG Public Consultation Paper on Draft Guidelines of Good Practice on Regulatory Aspects of Smart Metering for Electricity and Gas. [Accessed online on 1st March 2019]. Retrieved from http://www.smartgrids-cre.fr/media/documents/regulation/100326_ERGEG_Public_consultation.pdf

Energywatch. (2005). Get smart: Bringing meters into the 21st century. [Accessed online on 1st March 2019]. Retrieved from http://www.energywatch.org.uk/uploads/Smart_meters.pdf

Guerreiro, J., & Ferreira, L. M. (2015). Smart meter integration in an electric power system with large renewable energy resources, DSM and energy storage. [Accessed online on 1st March 2019]. Retrieved from https://fenix.tecnico.ulisboa.pt/downloadFile/395146453925/Paper_56622.pdf

Hennessy, M., & Keane, D. (1989). Lifeline rates in California: Pricing electricity to attain social goals. *Evaluation Review, 13*(2), 123–140.

Henney, A. (2013). Smart metering—A case study in Whitehall incompetence. New Power.

Hess, D. J. (2014). Smart meters and public acceptance: Comparative analysis and governance implications. *Health, Risk & Society, 16*(3), 243–258.

House of Commons Energy and Climate Change Committee. (2015). Smart meters: Progress or delay? 9th report of Session 2014–15, p. 18.

Ipsos MORI. (2017). Smart meter customer experience study. [Accessed online on 1st March 2019]. Retrieved from https://www.gov.uk/government/publications/smart-meter-customer-experience-study-2016-18

López-Rodríguez, M. A., Santiago, I., Trillo-Montero, D., Torriti, J., & Moreno-Munoz, A. (2013). Analysis and modeling of active occupancy of the residential sector in Spain: An indicator of residential electricity consumption. *Energy Policy, 62*, 742–751.

MacDonald, M. (2007). Appraisal of costs & benefits of smart meter roll out options: Final Report. Report for Department of Business Enterprise and Regulatory Reform, London. [Accessed online on 1st March 2019]. Retrieved from http://docplayer.net/19598967-Appraisal-of-costs-benefits-of-smart-meter-roll-out-options.html

Moore, A. E. (1935). The history and development of the integrating electricity meter. *Journal of the Institution of Electrical Engineers, 77*(468), 851–859.

National Audit Office. (2018). Rolling out smart meters. [Accessed online on 1st March 2019]. Retrieved from https://www.nao.org.uk/wpcontent/uploads/2018/11/Rolling-out-smart-meters.pdf

Ofgem. (2006). Domestic metering innovation. [Accessed online on 1st March 2019]. Retrieved from https://www.ofgem.gov.uk/sites/default/files/docs/2006/06/14527-metering-innovation_decision-document_final_0.pdf

Ofgem. (2019). Consumer vulnerability strategy. [Accessed online on 1st March 2019]. Retrieved from https://www.ofgem.gov.uk/system/files/docs/2019/10/consumer_vulnerability_strategy_2025_.pdf

Pepermans, G. (2014). Valuing smart meters. *Energy Economics, 45*, 280–294.

Pye, S., Dobbins, A., Baffert, C., Brajković, J., Deane, P., & De Miglio, R. (2017). Energy poverty across the EU: Analysis of policies and measures. In *Europe's energy transition-insights for policy making* (pp. 261–280). Cambridge: Elsevier Academic Press.

Santiago, I., Lopez-Rodriguez, M. A., Trillo-Montero, D., Torriti, J., & Moreno-Munoz, A. (2014). Activities related with electricity consumption in the Spanish residential sector: Variations between days of the week, autonomous communities and size of towns. *Energy and Buildings, 79*, 84–97.

Schultz, D., & Linewber, D. C. (2006). Real mass market customers react to real time differentiated rates: What choices do they make, and why? In *Association of Energy Service Professions Conference* (pp. 6–8). San Diego, CA.

Shove, E. (2015). No more meters? Let's make energy a service, not a commodity. [Accessed online on 1st March 2019]. Retrieved from http://theconversation.com/no-more-meters-lets-make-energy-a-service-not-a-commodity-40207

Torriti, J. (2010). Impact Assessment and the liberalisation of the EU energy markets: Evidence based policy-making or policy based evidence-making? *Journal of Common Market Studies, 48*(4), 1065–1081.

Torriti, J. (2012). Price-based demand side management: Assessing the impacts of time-of-use tariffs on residential electricity demand and peak shifting in Northern Italy. *Energy, 44*(1). 576–583.

Torriti, J. (2015) *Peak energy demand and demand side response. Routledge Explorations in Environmental Studies* (p. 172). Abingdon: Routledge.

Torriti, J. (2019). Flexibility. In Rinkinen, J., Shove, E., & Torriti, J. (eds.), *Energy fables: Challenging ideas in the energy sector* (pp. 103–111). Earthscan: Routledge

Torriti, J., & Grunewald, P. (2014). Demand side response: Patterns in Europe and future policy perspectives under capacity mechanisms. *Economics of Energy & Environmental Policy, 3*(1), 87–105.

Torriti, J., Leach, M., & Devine-Wright, P. (2011). Demand side participation: Price constraints, technical limits and behavioural risks. In Jamasb, T., & Pollitt, M. (eds.), *The future of electricity demand: Customers, citizens and loads* (pp. 88–105). Department of Applied Economics Occasional Papers (69) Cambridge: Cambridge University Press.

UKERC. (2017). The costs and impacts of intermittency – 2016 update, February 2017. [Accessed online on 1st March 2019]. Retrieved from http://www.ukerc.ac.uk/publications/the-costs-and-impacts-of-intermittency-2016-update.html

Watson, B. (2010). Smart and advanced metering in the UK – an overview. [Accessed online on 1st March 2019]. Retrieved from www.eatechnology.com

3   Cost–benefit analyses of smart meters in European countries

Background: the importance of cost–benefit analyses on smart meters

Several developed and developing countries are on the verge of taking important decisions on smart-metering implementation and have carried out various cost–benefit analyses (CBA) which measure the economic impacts of this radical change in electricity systems. The use of economic appraisals is understandable given that decisions around large public investments in smart-metering infrastructure are very difficult for policy-makers without the support of strong empirical evidence (Torriti, 2007). The decision on whether and how to spend several millions of Euros on such a radical change for electricity systems cannot be based solely on perceptions, lobbying and private interests.

European Union (EU) member states have an obligation to carry out a CBA as a result of the EU Third Energy Package in order to determine the viability of a smart roll-out. The recommended methodological approach to developing a CBA on smart meters is set out in the European Commission's Recommendation 2012/148/EU. Key aspects to the methodology set out in involve the approach; minimum functionality of the smart-metering system, costs and benefits.

Besides the EU obligation, carrying out an updated CBA will inform the complex decision-making around a very large investment for any EU government. The impacts from a roll-out of smart meters will involve a range of actors: (i) consumers (in terms of accurate costs, real-time information to enable them to manage energy consumption and potentially receive new services), (ii) Distribution System Operators (DSOs; in terms of management of frequent data on consumption and changes costs to serve) and (iii) society (in terms of carbon emissions). There are also benefits for the Transmission System Operator from the use, subject to appropriate data, privacy and access controls, of data collected through smart metering to better manage the network and to inform long-term investment in the electricity grid. Assessing the costs and benefits of intervention also means taking into account which volumes of smart-metering implementation would occur without direct government investment.

DSOs regularly invest large sums in utility equipment devoted to public service in pursuit of their regulatory or charter obligations to serve. Extending and maintaining service into deprived or underdeveloped distribution networks, for example, is a generally accepted unit of cost which is often implicit in the regulatory obligations. DSOs are well placed to assess the costs and benefits of these investments as they routinely fulfil these non-discretionary obligations while minimising the cost of doing so. Regulators are also prepared to defend their decisions in this cost-minimisation framework whilst relying on data supplied by utilities. However, smart-metering development may not fit into this time-tested paradigm of cost minimisation because it brings about a higher level of innovation and relates to the whole network. For instance, if a smart meter in a low-voltage distribution network can improve reliability beyond currently acceptability levels, it is unclear whether it is mandatory to invest resources to do so. This may depend on how much must be invested to obtain the improvements for DSOs, suppliers and consumers and whether the improvement gained is worth the money.

Performing CBA on smart meters poses interesting and challenging problems in measuring physical impacts and estimating economic benefits from them. However, when smart meters are part of first-of-kind or pilot projects, there are additional challenges to producing meaningful CBA. These complexities call for a consistent approach in selecting the type of data and methodologies which should be included in any CBA on smart-metering implementation.

This chapter summarises the data on costs and benefits available from official documents on CBA of smart meters.

Cost-benefit analyses of smart meters in European countries

This section presents the main features of existing CBAs of smart metering in the following European countries: Germany, the United Kingdom, the Netherlands, Ireland, Hungary and France. In this country-by-country analysis the main emphasis is on baseline scenarios (i.e. Business as Usual without smart meters) and 80% smart-metering implementation by 2020 (often referred to as the "EU scenario").

It concludes by identifying key lessons learned for other countries in terms of data and methodological issues associated with categories of costs and benefits.

Germany

In absolute terms, Germany has the highest annual electricity demand in Europe, the largest generation portfolio and its power system interconnected with ten countries. Over the past decade, Germany experienced a significant transition to renewable energies, a phenomenon also known as *Energiewende*. This transition called for the deployment of smart meters.

The German CBA on smart meters considers various scenarios for smart meter roll-out. This section of the chapter focuses on the "EU scenario" (Ernst & Young, 2013). The "EU scenario" is most consistent with the requirement for an 80% roll-out of smart metering by 2022.

The technological choice of Germany consisted of a gateway located in the consumer premises that manages the transfer of data to necessary parties. It provides high levels of data protection (privacy and security) to the customer. Two important considerations are that the gateway configuration involves higher costs than other approaches and the benefit of allowing the connection of other utility services to the same infrastructure (e.g. gas and heating).

Table 3.1 shows the main assumptions in the CBA. The benefits and costs are assessed and discounted for a period of 20 years (from 2012 to 2032). The main assumption regards changes in consumption associated with the presence of smart meters in end users' premises. These vary between 0.5% and 2.5%. This is a five-fold variation which can distort significantly the net impacts of smart meters in Germany.

Figure 3.1 shows the discounted costs of smart meters, IT, communications, in-home display (IHD) and training. Although the CBA does not provide a full breakdown of costs and benefits of the EU roll-out scenario, there are assessments of costs and volumes. Hence, in Figure 3.1 the values are expressed in terms of Euro per smart meter. The average cost of communications is particularly high when compared with other CBAs. This is because the communications configuration which Germany opted for consists of 80% use of GPRS and only 20% use of PLC. Hence, the high proportion of GPRS is a key driver of overall communications costs. This is considered necessary to permit remote control of various devices (for instance, solar panels) and not just for remote meter reading. Overall, costs are broken down between capital and operating expenditure, with 38.5

Table 3.1 Main assumptions underpinning German cost–benefit analysis

Key assumption	Unit	Value
Roll-out period	Years	2012–2022
Proportion of metering points covered	%	80% by 2022
Modelling period	Years	2012–2032
Discount rate	%	5% commercial, 3.1% residential and industrial
Asset life of meters	Years	13
Number of avoided meter readings	Number/meter	1
Reduction in consumption	%	Between 0.5% and 2.5%
Peak load transfer	GW	6.1
Reduction in non-supplied energy	%	1%
Reduction in theft	%	20%

million electricity smart meters installed over eight years. The installation of smart meters reflects growth in customer numbers.

Figure 3.2 lists the estimated benefits in the German CBA. The highest benefits consist of electricity cost savings. These are analysed in more detail below. Reduction in meter readings and avoided investment in standard meters are also large sources of benefits. Indeed, the average billing costs for heating are estimated to be around €80 per year, which could offset significantly any additional gateway costs.

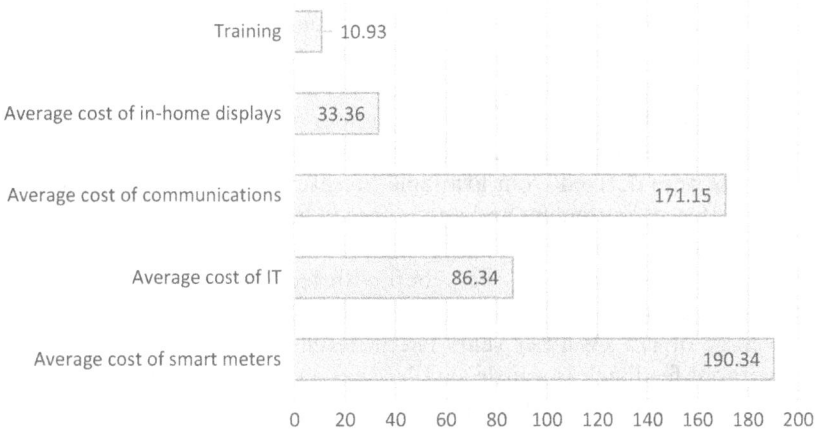

Figure 3.1 Discounted costs (€/metering point).

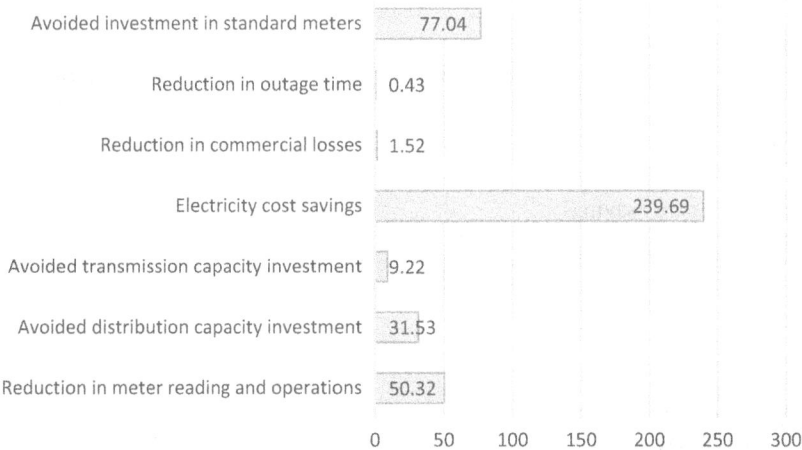

Figure 3.2 Discounted benefits (€/metering point).

Overall, this assessment shows not only high costs but also high benefits. The "EU scenario" is associated with a Net Present Value (NPV) of −€0.1 billion. Hence, overall the CBA presents marginally negative results.

The sensitivity analysis shows a significantly negative NPV (−€5.9 billion) for zero net conservation effect and a positive NPV of +€6.1 billion for a 3.6% net conservation effect. The high volatility in the results depends mainly on the net conservation effect. The sensitivity analysis on the number of meter operators shows a positive NPV (€0.6 billion) if the current 900 smart meter providers are amalgamated into 70 and a NPV rising to €1.1 billion in the case of there being ten meter operators. This finding suggests significant economies of scale in the development costs of IT infrastructure.

Since net conservation effect and, consequently, energy savings are the most significant source of benefits for the German CBA, these are looked at here in closer detail. Chapter 6 will then discuss net conservation effects in relation with other studies. Table 3.2 summarises different levels of energy savings. The starting points consist of saving percentages. Different levels of savings were derived from available literature. Only knowing the influence of different equipment and appliances in households could a statement about the saving effects of feedback be made. A previous study in Germany by Fraunhofer showed that the group with feedback had an energy consumption which was 3.7% lower than the reference group. This amounts to an average of 125 kWh per year. The difference between the groups with and without feedback is a little smaller than expected – older studies from other countries found savings of 7%, although the circumstances were very different there. In Linz, the net conservation effect remained at the same level during the whole field trial, whereas in Germany no reliable statement about the stability of the effects can be made as the data sample was insufficiently large.

Table 3.3 illustrates the findings from the sensitivity analysis with regard to those changes which induce a negative variation in NPV. The highest sensitivity is associated with energy savings. This means that a nil effect of smart meters on end users would reduce the societal benefits of smart meters to an extent that the NPV would decrease by €5.7 billion. A shortfall in terms of grid efficiency would also generate a substantial reduction in

Table 3.2 Energy savings

Consumption	Savings (%)	Load shifting (%)	Average cost savings per meter (€ per year)
<2,000 kW/a	−0.5	0.25−5	2.5
2,000–3,000 kW/a	−1	0.5−10	10
3,000–4,000 kW/a	−1.5	0.75−15	20
4,000–6,000 kW/a	−2	1.0−20	39
>6,000 kW/a	−2.5	1.25−25	75

Table 3.3 Sensitivity analysis: changes which induce negative net present values

Changes/sensitivities	Variations in NPV (billions of €)
No energy savings	−5.7
Shortfall of grid efficiency	−2.9
Extension of the deadline to smart meters from 2018 to 2022	−0.7
Periodic replacement after 24 years	−0.6
Shortfall of economies of scale of procurement	−2.2

benefits (€2.9 billion). Extension of the deadline to smart meters from 2018 to 2022, periodic replacement after 24 years and shortfall of economies of scale of procurement were also considered as sensitivities bringing about negative changes to the overall profitability of smart-metering implementation.

United Kingdom

The UK government took the decision to mandate a roll-out of meters in October 2009, a year and a half before the results of the pilot trials were published. The UK roll-out is characterised by its voluntary nature and by the fact that it is supplier-led.

The UK government produced several CBAs on smart meters. The first one was commissioned over ten years ago (MacDonald, 2007), whereas the most recent was produced by the Department of Energy and Climate Change (DECC) in 2016. The main reason for the multiplicity of CBAs is that the UK government revised timetables under which suppliers are required to take all reasonable steps to complete their roll-outs by the end of 2020. These developments were reflected in the different version of Impact Assessments published between 2009 and 2016.

In between these assessments, the UK government conducted other Impact Assessments. In 2009, an Impact Assessment informed the appraisal of alternative options for the preferred market model for the roll-out. Options previously considered and discarded include a fully competitive model, a fully centralised model and a DNO-led deployment.

In 2010 and 2011 government considered options for the implementation of the preferred market model: a supplier-led roll-out with centralised provision of communications and data services. Detailed policy design options were considered and assessed. These included the completion date, the establishment and scope of Data and Distribution Company (DCC), the functionality of the smart meter, the roll-out strategy and the strategy for consumer engagement.

The 2011 Impact Assessment set out the government's overall approach and timeline for achieving smart-metering roll-out. In 2012, the Impact Assessment was further updated with additional evidence base and supported the introduction of the first tranches of smart-metering regulations.

The economic case for smart meters in the United Kingdom improved dramatically in about a decade. The initial assessment in 2007 stated the following: "Provision of feedback through advanced metering solutions is heavily burdened by the high costs associated with legacy meters and developing the communications infrastructure".

Remarkably, the Net Present Value of smart-metering implementation was estimated at the time to be −£4 billion. This raised up to over £6 billion in the most recent CBAs, as detailed below.

The Impact Assessment carried out by DECC presents a largely positive Net Present Value of £6.2 billion for the preferred policy option, which consists of carrying on with the mandatory roll-out of smart meters and setting out additional regulations associated with their implementation.

The overall findings of the CBA (i.e. the economic assessment underpinning the Impact Assessment) are as follows.

With regard to costs, IHDs, meter and its installation and operation amount to £6.36 billion. DCC-related costs, including communications hubs provision, amount to £2.47 billion. Energy suppliers' and other industry's IT systems costs amount to £0.79 billion. Industry governance, organisational and administration costs, energy, pavement reading inefficiency and other costs amount to £1.30 billion.

Figure 3.3 shows the percentages and absolute spending associated with different typologies of costs. Capex and opex costs are the highest category, followed by communications costs. Suppliers were expected to face the following costs: capital costs of smart meters, IHDs and potentially the communications hub that links the meter(s) in a property to the supplier via the DCC and installation, operation and maintenance of this

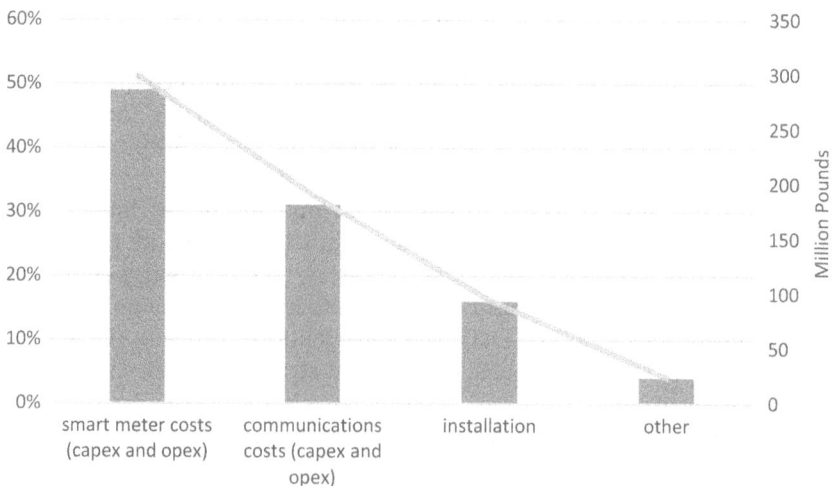

Figure 3.3 Typologies of costs (percentages and absolute spending).

equipment. Consumers are connected with allocation of upfront investment in supporting IT systems and the DCC, as well as their ongoing maintenance. Distribution Network Operators are also foreseen to incur costs in order to upgrade their systems in order to integrate into the smart meter network.

With regard to benefits, total consumer benefits amount to £5.73 billion and include savings from reduced energy consumption (£5.69 billion) and microgeneration (£36 million). Total supplier benefits amount to £8.26 billion and include – amongst others – avoided site visits (£2.97 billion) and reduced inquiries and customer overheads (£1.19 billion). Total network benefits amount to £0.99 billion and generation benefits to £85 million. Carbon-related benefits amount to £1.21 billion. Air quality improvements amount to £95 million.

Figure 3.4 illustrates the percentage and absolute amounts associated with different typologies of benefits. Energy savings for end users feature as the highest share of benefits. Suppliers are also expected to experience benefits from fewer meter reads, theft reductions and reduced suppliers' call centre traffic. Network operators are expected to improve electricity outage management and resolve any network failures more efficiently. However, smart grid benefits were not quantified in the CBA.

The main uniqueness of CBAs on smart meters in the United Kingdom are that the main stakeholders involved are not the DSOs, but the suppliers. A key feature of the CBA in Great Britain is that it predicts a dual fuel electricity and gas smart meter roll-out. A CBA of this nature reflects the fact that the metering market is de-regulated and that many customers use the same provider for gas and electricity suppliers, and hence in the British

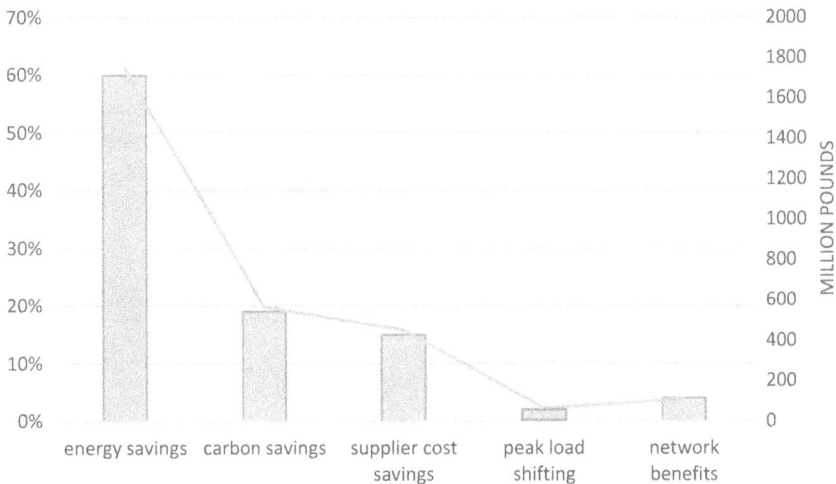

Figure 3.4 Typologies of benefits (percentages and absolute amount).

supplier-led model there might be strong incentives to replace both electricity and gas meters with smart meters at the same time. This dynamic is rarely experienced by the other countries reviewed in this chapter.

The different Impact Assessments present dissimilar Net Present Values. In 2014 the NPV increased by £184 million in comparison to the IA published in August 2011.

Table 3.4 presents the different findings of the CBAs performed between 2011 and 2014. Over the years the CBA was updated to account for the changes in assumptions and projections on fossil fuel prices, carbon prices, carbon emission factors, energy consumption and number of meters in both domestic and non-domestic sectors. With regard to non-domestic CBAs, both total costs and benefit estimates increased over time. This is mainly a result of changes in carbon and energy prices as well as the move of the present value base year. The move of PV base year results in increase in both costs and benefits, but with the benefit increase having a stronger impact. Updated planning and roll-out profiles as well as cost uplift to early meters slightly counteract this effect. Updated planning and roll-out profiles resulted in fewer meters expected to be in place by the end of 2014 than previously modelled, resulting in some benefits and costs from smart meters occurring slightly later in time.

The Netherlands

Historically, the main sources of electricity in the Netherlands have been fossil fuels (e.g. natural gas and coal). More recently, renewable energy sources (e.g. wind energy and solar energy) generated an increasing share of electricity. Consumption grew with an average of 4.5% per year over the last six decades (IEA, 2017).

The CBA was carried out on the draft bill amending the Electricity Act of 1998. Key features of this bill amendment consist of making metering a regulated activity by introducing regulated tariffs; placing an obligation on DSOs to undertake a nationwide roll-out of smart meters by over a six-year period for all small customers; and making DSOs responsible for the provision and maintenance of meter infrastructure and making suppliers

Table 3.4 Changes in results of cost–benefit analyses on smart metering carried out by UK governments between 2011 and 2014 (million pounds)

	NPV	*Total costs*	*Total benefits*	*NPV difference*
2011 CBA non-domestic	£2,154	£604	£2,759	0
2012 CBA non-domestic	£2,338	£608	£2,946	£187
2013 CBA (all end users)	£6,659	£12,115	£18,774	0
2014 CBA (all end users)	£6,214	£10,927	£17,141	£445

responsible for collecting and processing of raw meter data and communicating it to customers (CBS, 2016).

With regard to the CBA, the starting points were: the benefits of smart metering might be mostly on suppliers and consumers, whereas costs are borne by DSOs; the CBA focused exclusively on the impact of smart meters on DSOs; and large areas of DSO benefits other than metering are excluded (such as network management, impact on demand forecast accuracy, impact on annual energy consumption, impact on peak demand).

The CBA considers costs which could be recovered through the metering tariff. This includes the potential cost of stranded traditional metering assets and the cost of ongoing manual data collection during the period of the rollout.

In terms of scope, the CBA applies to those sites that only have electricity supply (about 845,000 customers). Costs differ depending on whether the electricity supplier chooses to use its priority privileges or not.

Figure 3.5 shows site-specific and meter-specific costs in the Dutch CBA. Collection of input data on costs and benefits involved a review of published data sources, and bilateral discussion with Dutch market participants and meter manufacturers. Purchase costs, installation of new meters and GPRS communication costs feature as the highest costs.

Figure 3.6 shows that DSOs are assumed to be able to save between EUR 4.40 and EUR 6.50 per meter per annum, because of the combined impact of network management benefits and cost savings achieved, thanks to the transfer of responsibility regarding data collection and validation.

The review of costs and benefits was used to generate a range for each element of costs and benefits. An economic model was developed to estimate

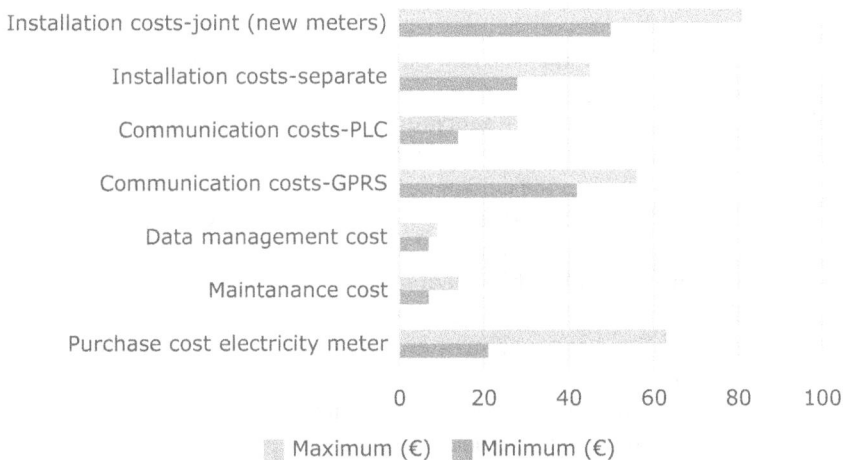

Figure 3.5 Costs of smart meters (in € per meter, from international review).

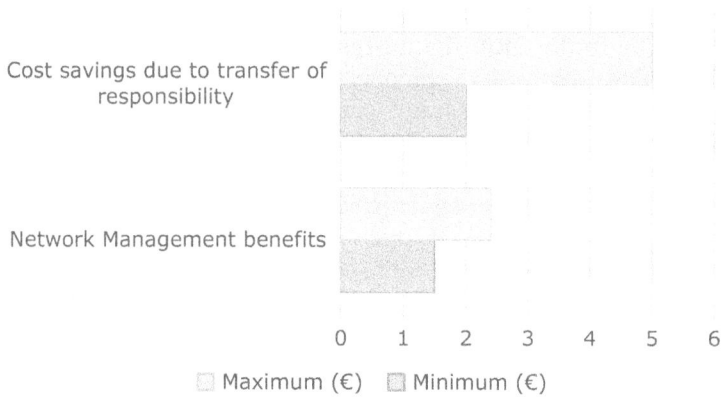

Figure 3.6 DSO annual costs (in € per meter, from international review).

the likely level of net cost relative to the tariffs. The model was used to assess the impacts of smart meter roll-out on DSOs.

One of the most relevant features of this CBA is that direct feedback is not included as a category of benefits. It is assumed instead that bi-monthly readings are provided to end-users, with additional information on electricity consumption. In addition, several benefits in Recommendation 2012/148/ EU are not included in the CBA. For instance, deferred investment, technical losses, CO_2 emissions and air pollution are not listed as benefits in the analysis. An additional benefit consists of increased competitiveness following smart-metering installation. The benefit is accounted as an allowance associated with the ability for competitors to develop new market niches in a market with a full roll-out of smart meters.

For the first five years of the smart-metering roll-out, the net benefits are negative. By the end of the roll-out, net benefits become positive. This is a consideration which does not apply to the Dutch case only, as smart-metering roll-out periods tend to be financially straining for main implementers. However, in CBA the mere comparison of undiscounted benefits and costs is often not as significant as the calculation of the Net Present Value, following the application of a discount rates on both future benefits and costs.

A discount rate of 5.5% per annum is applied to the future stream of costs and benefits in the CBA. This is equal to the Weighted Average Cost of Capital (WACC) used by Energiekamer in their regulation. The discounting is operated over 17 years based on average technical and economic lifetime of smart meter. Two different cumulative Net Present Values were calculated under the optimistic and pessimistic scenarios used in the Dutch CBA. In the optimistic scenario, the net benefits break even after nine years. In the pessimistic scenario, the NPV stays negative for entire duration of the assessment.

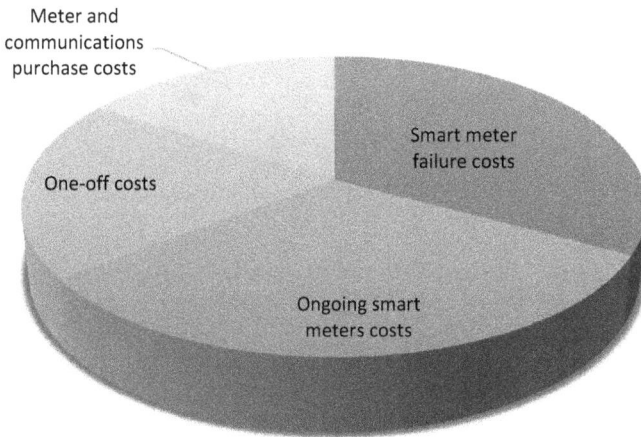

Figure 3.7 Sensitivity analysis of main categories of costs and benefits.

Figure 3.7 shows the sensitivity analysis of main categories of costs and benefits. The sensitivity analysis emphasises uncertainty about smart-metering failure costs (higher failure rate than legacy meters). Similarly, the sensitivity analysis points to uncertainty especially about DSO communications costs (i.e. PLC vis-à-vis GPRS).

In the sensitivity analysis, a key item consists of competitiveness. Removing such item would trigger negative net benefits (i.e. −€19 per metering point, equivalent to −€127 million in total).

Ireland

Electricity demand in Ireland has been decreasing since 2010, but in 2015 there was an increase by 2.9% in conjunction with economic growth. Ireland's smart meter rollout involves about 2.2 million electricity consumers with an investment of up to €1 billion. It is expected to yield a net benefit of around €229 million over 20 years (SEAI, 2015).

Figure 3.8 illustrates the assumed roll-out schedule in Ireland. In the CBA, it is assumed that electricity smart meters are rolled out on a constant basis between 2017 and 2020 to private residences in Ireland as well as to Small and Medium-sized Enterprises (SMEs). It is also assumed that IHDs are provided in private residences only and are installed with the smart meter. The CBA is based on the principle that time of use tariffs are introduced to private residences and SMEs one year after installation of the smart meter.

With regard to categories of consumer costs and benefits, smart meter learning time is considered as a cost, whereas reduced consumption costs and ongoing time savings are types of benefits.

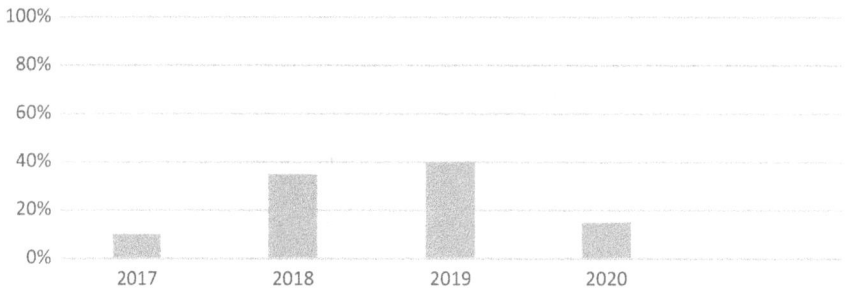

Figure 3.8 Assumed roll-out schedule.

Table 3.5 highlights how for a residential household with smart meters, IHDs and time of use tariffs, the average annual saving on an electricity bill is €16.2. This saving results from an overall reduction in energy consumption (i.e. 134 Kwh per annum) and from a reduction associated with peak hours following time of use tariffs incentivising off-peak consumption over peak-time demand.

These savings should take place in full in the year following smart metering, IHDs and time of use tariffs implementation. The CBA mentions international evidence in terms of successful cases of energy savings.

The CBA focuses on costs and benefits for the Electricity Supply Board (ESB), which historically acted as and now operates as a commercial semi-state organisation in a liberalised and competitive market. With regard to smart meters, ESB assumes a significant responsibility for the roll-out as well as for their management (extending to data flows and storage) once implementation has taken place. The CBA identifies significant associated capital costs beyond the costs of the SMs and IHDs. These are in terms of programme management (i.e. the costs which ESB will incur for the management of the full national roll-out of the programme, including procurement); head end system (i.e. the software needed to ensure a two-way communication between the smart meter and the ESBN meter data management system); meter data management system (i.e. the system needed for the storage, management and distribution of consumption data from the smart meters); backend enhancement (i.e. the current system used by ESBN for meter management); deployment and materials management; investment in systems to ensure an effective management of the programme of smart meter roll-out; and IT security (i.e. investment required in new applications to ensure that smart meter data is secure at all points of transit and processing). In addition, the CBA points to operating costs for ESBN associated with smart-metering implementation, consisting of mobile operator charges (i.e. charges for ongoing communication between smart meters and the meter data management system of the ESBN); telecoms operation

Table 3.5 Profile of usage-related benefits

	Pre-smart meter			Post smart meter		
	Consumption	Average unit cost	Total cost	Consumption	Average unit cost	Total cost
Peak	433	0.11	47.5	390	0.15	60.5
Shoulder	3,210	0.11	349.4	3,111	0.11	334.8
Off-peak	929	0.11	99.5	937	0.09	85
Total	4,573	0.11	496.5	4439	9.24	480.3

and maintenance (i.e. the costs associated with operating and maintaining the telecoms infrastructure required to support smart meter to meter data management system communications); head end and meter data management system annual management and enhancement (i.e. the annual costs associated with the operation and maintenance of these systems, as well as their occasional enhancement); Network Operations Centre (i.e. the costs of the team charged with ensuring that smart meter business processes are fit-for-purpose); data storage costs (i.e. the costs of storing the data from the smart meters as they are rolled out); replacement of faulty meters (i.e. the costs associated with the replacement of faulty smart meters); IHD-related costs (i.e. back-office and customer costs during the planning and roll-out phases, communication costs during the two-year management period and faulty IHD replacement).

With regard to benefits for ESBN, these consist of avoided manual meter roads; avoided costs of meter replacement; avoided costs of new connections; avoided costs of specialist pre-payment meters; avoided support costs for prepayment meters; avoided costs of microgeneration capability on traditional meters; avoided visits to traditional meters (e.g. disconnections or reconnections); avoided costs associated with investigation of voltage complaints; avoided costs of re-enforcing the network as a result of reduction in peak demand and as a result of overall reduction in consumption; and avoided theft. Specifically, the demand-related benefits consist of €1.04 million for every 1% reduction in energy consumption and €0.73 million for every 1% reduction in peak energy consumption.

Figure 3.9 shows suppliers' capital costs associated with smart metering. Enhancing the Customer Relationship Management (CRM) system and the billing system represent the highest costs for suppliers. These estimates on capital costs are based on the fact that all suppliers were interviewed regarding impacts of smart-metering implementation on their financials. Only three suppliers provided comprehensive responses. In addition to capital costs, opex cost categories consist of staff training, IT opex, customer awareness raising, smart bill production and printing of shadow statements, additional communications costs.

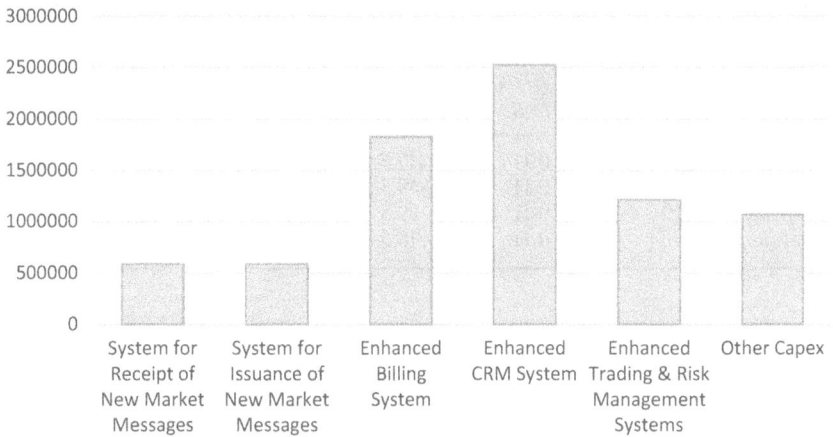

Figure 3.9 Suppliers' capital costs of smart metering.

With regard to suppliers' benefits, the main categories consist of avoided ad hoc meter reads, bad debt reduction, and reduction in dunning cycles.

Table 3.6 shows the benefits and costs of changes brought about by smart meters with regard to electricity generation. Reductions in peak demand bring about cost saving, thanks to avoided generation. The NPV of this benefit is €68 million.

Table 3.7 shows the benefits associated with lower system marginal prices as a result of lower demand. This is caused by the assumption that a fall in residential demand brings about a fall in wholesale price which leads to lower system marginal prices.

Figure 3.10 presents the full CBA findings with different Net Present Values for all stakeholders. ESBN and electricity suppliers face negative NPV values. The scale of the net cost in the case of networks is very substantial and reflects in large part the assumed national use of a GPRS communications solution with a high attendant annual operating cost. With regard to consumers, residential households are associated with positive Net Present Values in aggregate, while SMEs' Net Present Values are negative.

Figure 3.11 shows the sensitivity of critical variables to changes in assumptions. The CBA turns positive in terms of Net Present Value either when consumers build on their initial experience of time of use tariffs and derive further usage benefits from their smart meters, IHDs and time of use tariffs over time or when estimated costs are 10% lower than anticipated. Benefits compounding are experienced when major categories of cost are 10% lower than initially estimated, and consumer usage benefits compound gradually over time to reach 175% of the initial Cost Benefit Analysis assumption by 2033. This benefit compounding is a standard feature of many other smart-metering CBAs reviewed in this chapter.

Table 3.6 Electricity generation benefits and costs

Type of change	Benefit/cost
Return on investment residential capacity pot (in MW)	1,869
Average cost per MW	€78,730
Reduction in peak demand	−9%
Benefit deflation factor	32%

Table 3.7 Benefits associated with lower system marginal prices

Value of total, residential, ROI smart metering in 2011	€1,730,510,620
Percentage reduction achieved	0.40%

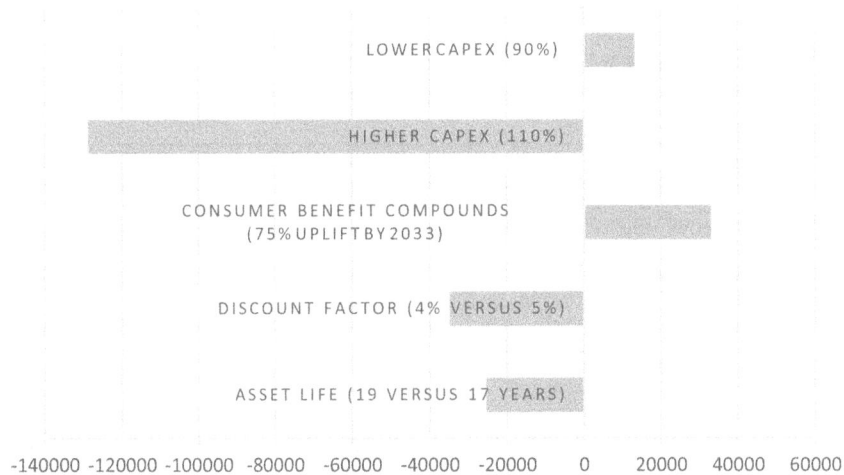

Figure 3.10 Sensitivities of different variables (in thousands of Euros).

The extent to which the total of the loss is reduced (i.e. less negative Net Present Value in absolute terms) depends on both changes in the assumed duration of the smart meter asset life and when the discount factor is reduced to 4%.

On the other hand, a capex higher than forecast by 10% results in a more negative Net Present Value than for the Cost-Benefit Analysis results.

Hungary

Hungary has an electricity consumption of 4.1 MWh per capita. During the period covering 2005–2015, Hungary experienced a double-digit reduction

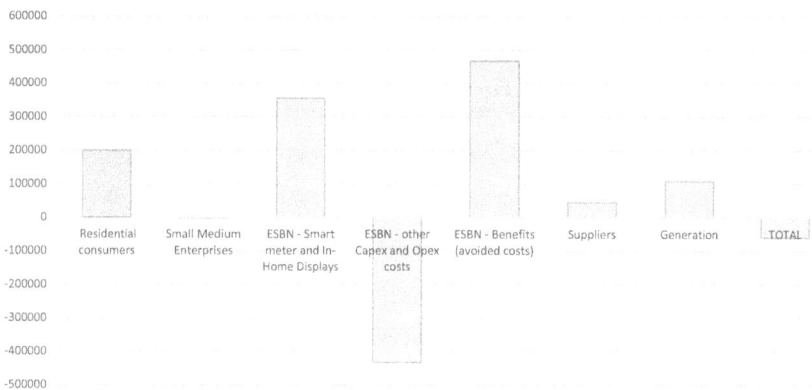

Figure 3.11 Net Present Values for different stakeholders (in thousands of Euros).

Table 3.8 Key assumptions in Hungarian cost–benefit analysis

Key assumption	Value
Discount rate	10.30%
Asset life of meters	15 years
Number of avoided meter readings	1
Reduction in consumption	1.50%
Reduction in theft	50%
Reduction in technical losses	1.50%

in net electricity generation, compared with average 2.6% reduction in the level of EU-28. During the same period, the market share of the largest electricity generator rose from 38.7% to 53.1% (Eurostat, 2018). Hungary is one of the European countries with the lowest electricity consumption per person – the lowest of the EU-28 countries in 2012, for instance (IEA, 2018). In the current legal framework, the DSOs are responsible for the installation and maintenance of the meters.

Table 3.8 illustrates the key assumptions in the Hungarian CBA. The theft reduction figure is particularly high when compared with other CBAs. The approach adopted in the Hungarian CBA is based on aggregate annual cost figures, with benefits estimated as proportional decreases from the revenue bases.

The Hungarian roll-out involves a plan for 80% installation by 2021. The alternatives considered in the CBA involve a joint roll-out by all DSOs, in which the companies jointly develop the system that processes and stores the data provided by smart meters; and the Transmission System Operator taking over the functions of the Smart Meter Operator.

Like for other CBAs, the demand management functions are not included as part of the benefits of the smart meter set up.

When estimating costs and benefits over time a set of assumptions were considered. The number of manual meter readings saved per year was 1. The average billing cost per customer with smart meters was €0.50 per metering point per year. The customer care cost/customer/year baseline was €2 per metering point per year. Peak load transfer consists of 2%. The percentage of customers requesting incremental contracted power is 1%.

Figure 3.12 shows the cumulative phasing-out of standard meters and increasing penetration figures for smart meters.

Figure 3.13 shows the overall findings of the Hungarian CBA. The net benefits are −78.20 Euros per metering point.

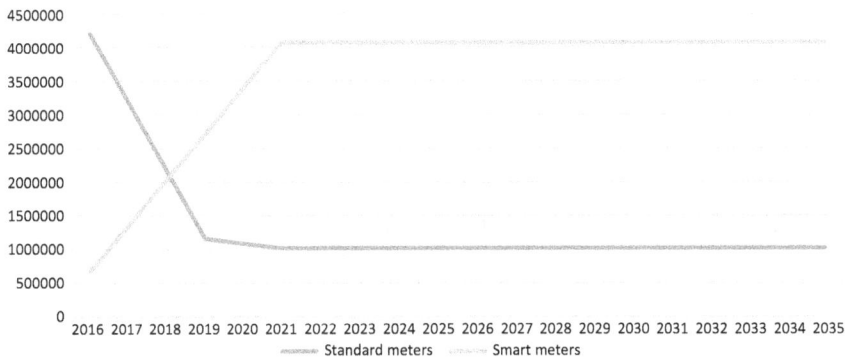

Figure 3.12 Cumulative penetration of smart meters and standard meters.

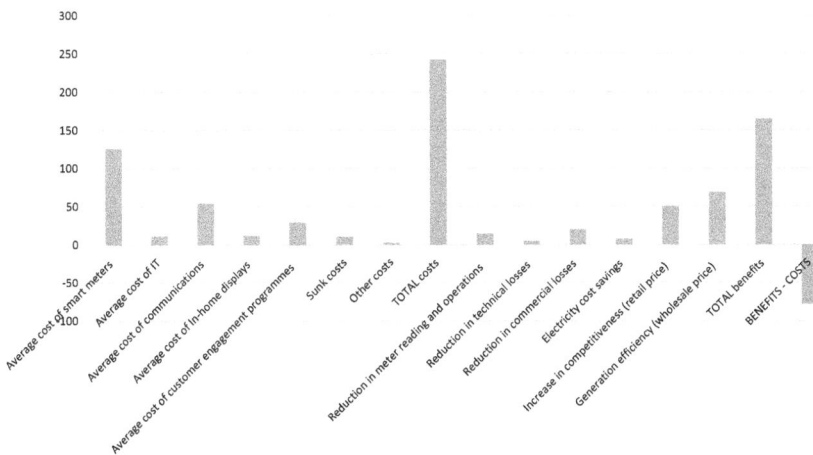

Figure 3.13 Benefits and costs in Hungarian CBA (€ per metering point).

The most optimistic scenario comprises a 10% reduction in network losses; a 75% reduction in customer switchover costs; a 16% decrease in balancing power demand; and a 6% decrease in wholesale prices, relating to a 2% reduction in end-user prices due to more intense competition.

France

France has an average electricity consumption of 7.04 MWh per capita. France has 35 million electricity meters in place and ambitious goals for the development of renewable source of electricity and energy efficiency, together with a commitment to reduce the share of nuclear energy in the power mix. Electricity demand in France depends significantly on seasonality and is temperature-sensitive.

DSOs are legally responsible for all operations related to metering, including installation, maintenance and meter readings. Before smart-metering implementation, meters were read manually twice a year.

The first CBA was conducted by the Regulatory Commission of Energy in 2007. Successively, a large pilot study was conducted. The pilot study consisted of a two-year project with 300,000 customers in two regions. With regard to costs, the pilot project points to 60% from installation, 30% from purchase and 10% from other. From the pilot, it was estimated that a national roll-out would cost in the region of €4.3 billion. With regard to the distribution of benefits, it was estimated that 55% of benefits would consist of reductions of non-technical losses, 40% performance of interventions and 5% better asset management and network operation (RTE, 2016). Following the pilot project, a financial CBA was carried out. The costs and benefits are extremely consistent with the outcomes of the pilot project.

With regard to the distribution of costs and benefits, DSOs would experience losses as a result of smart-metering roll-out, more so if this took place over five years. The costs experienced by DSOs would consist of investment costs (including metering equipment, meter installation, hubs equipment and meter information system); stranded costs (i.e. replacement of the meter in advance); and operating costs (i.e. maintenance, repair and operations for meters and hubs, and operations information system). DSOs would also experience benefits in terms of remote reading and grid optimisation. It was estimated that a roll-out in ten years would reduce costs for DSOs by around 15%. These lower costs are not sufficient to generate positive Net Present Values.

Table 3.9 shows the overall costs of smart-metering implementation in France divided by cost of smart-metering units, information technology and communications costs.

Table 3.10 illustrates the main benefits. The highest benefits consist of avoided investment in installing existing meters, followed by avoided network losses and avoided meter reading costs. In addition to the material

Table 3.9 Costs in French cost-benefit analysis

Cost type	Billions of €
Smart meters	3
Information technology	0.5
Communications	0.3
Total	3.8

Table 3.10 Benefits in French cost–benefit analysis

Type of benefit	Benefit percentage (%)	Discounted benefits (in billions of Euros)
Avoided investment in installing existing meters	30	1.5
Avoided network losses	25	1.2 or 1.8
Avoided meter reading costs	15	0.7

presented in Tables 3.9 and 3.10, it was estimated that the operating costs of smart-metering systems would be €0.7 billion higher over the period between 2011 and 2038, but there would be other benefits in terms of operating expenses (€0.1 billion) and reduced technician's interventions (€1 billion). Hence, the Net Present Value varies between €0.1 billon and €0.7 billion, with the highest sensitivity around avoided network losses, which may vary between €1.2 and €1.8 billion.

Conclusions

In most of the European countries reviewed in this chapter the economic appraisals of smart meters reflect contrasting positions over their benefits and costs.

Overall, the CBAs in the European countries reviewed in this chapter point to significant lessons that can be learned for economic analysis on smart-metering implementation in other countries. The main lessons learned relate to data on reductions in electricity consumption and loss reductions.

With regard to consumption reduction, this is critical for CBA because the relatively small percentage benefits in terms of savings are multiplied by millions of users. Chapter 6 investigates this point in more depth with more a more comprehensive set of trials from different countries.

Smart meters have been shown to provide significant benefits to DSOs in countries with high levels of theft and commercial losses. Similarly, the

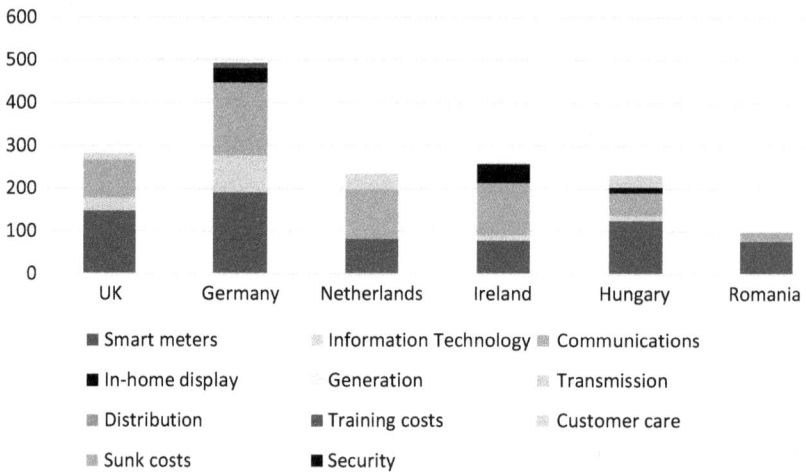

Figure 3.14 Costs in CBAs on smart-metering implementation (in Euros per meter, discounted).

scope to defer peak investment, reduce technical losses and reduce the time of outages will depend on the starting position of the country.

Figure 3.14 illustrates the discounted costs for the European countries reviewed in this chapter. The final cost of smart meters includes installation. Different labour costs in each country will affect this unit of cost. For instance, the experience of Italy and Spain, which purchased high quantities of smart meters, points to lower costs of the meters in tenders. The high variation in the units of costs in Figure 3.14 is due to IT and communications costs. There is no standard criterion for their cost allocation. In this regard, the graph below shows that the total cost of these two items is generally in a range between €50 and €100. In Germany, the cost for the entire system is estimated at over €233 per metering point. The most expensive communications technology (GPRS) is adopted in Germany. On the other hand, the cheapest communications technology (PLC) is principally applied in the countries with lowest costs, including Romania and Hungary. About 93% of costs are associated with communications, IT and meters costs. Overall, total discounted costs for the United Kingdom, Germany, the Netherlands, Ireland and Hungary are, respectively, €281.65, €492.12, €240.28, €260.49 and €242.42.

Figure 3.15 illustrates the discounted benefits per meter as estimated in the CBAs reviewed in this chapter. Reduction in meter reading operations varies from €14.5 (in Hungary) to €145.8 (in the United Kingdom). The discrepancies between reduction meter reading costs can be connected to divergent regulatory and operational arrangements with regard to the billing cycle and differences in labour costs (as explained above about

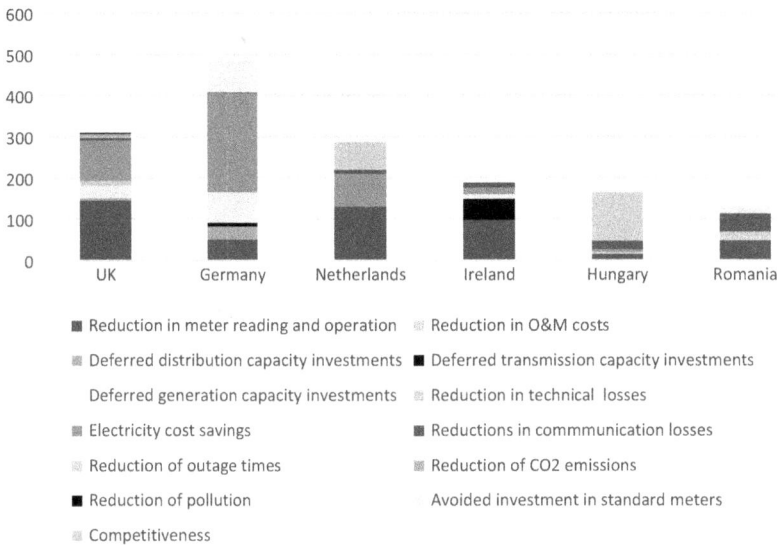

Figure 3.15 Benefits in CBAs on smart-metering implementation (in Euros per meter, discounted).

installation costs). Reductions in technical losses of electricity are high only for the United Kingdom and Romania. Overall, total discounted costs for the United Kingdom, Germany, the Netherlands, Ireland and Hungary are, respectively, €308.9, €484.9, €287.3, €187.7 and €164.6.

References

CBS. (2016). National energy outlook. https://www.cbs.nl/-/.../45/national%20energy%20outlook%202016_summary.pdf

Ernst & Young. (2013). Kosten-Nutzen-Analyse für einen flächendeckenden Einsatz intelligenter Zähler. Berlin: Author.

European Commission Recommendation 2012/148/EU of 9 March 2012 on preparations for the rollout of smart metering systems (OJ L 73, 13.3.2012, p. 9)

Eurostat. (2018). Electricity production, consumption and market overview. http://ec.europa.eu/eurostat/statistics-explained/index.php/Electricity_production,_consumption_and_market_overview#Main_statistical_findings

IEA. (2017). Netherlands – energy system overview. https://www.iea.org/media/countries/Netherlands.pdf

IEA. (2018). Electric power consumption (kWh per capita). http://www.iea.org/statistics/statisticssearch/

MacDonald, M. (2007). Appraisal of costs & benefits of smart meter roll out options. Final chapter. Chapter for Department of Business Enterprise and Regulatory Reform, London.

RTE. (2016). Production nationale annuelle par filière (2012 à 2016). https://opendata. rte-france.com/explore/dataset/prod_par_filiere/table/?sort=-annee

SEAI. (2015). Energy in Ireland 1990–2014. http://www.seai.ie/resources/publications/ Energy-in-Ireland-1990-2015.pdf.

Torriti, J. (2007). (Regulatory) Impact assessments in the EU: A tool for better regulation, less regulation or less bad regulation? *Journal of Risk Research, 10*(2), 239–276.

4 Cost–benefit analysis of smart meters in Romania

Overview of smart meters in Romania

Since the approval of the Common Rules for the Internal Market in Electricity Directive (2009/72/EC) in 2009, Romania has been attempting to assess the impacts of smart-metering implementation. The EU Directive aimed to achieve an 80% installation of electricity smart meters in Romania by 2020, if an economic assessment proved the implementation as feasible.

The main analytical effort to understand the impacts of smart-metering implementation in Romania was carried out as part of a study performed by AT Kearney in 2012. The analysis showed favourable results associated with the 80% roll-out of smart metering in Romania up to 2020. Following this positive economic assessment, Romania legally committed to the 80% implementation of smart meters by 2020.

The implementation of smart metering in Romania will have large economic, social and environmental impacts. The expenditure associated with smart meters may be in the region of between €700 million and €1 billion. Reduction in losses or energy consumption will save fuel and reduce carbon emissions.

Despite previous efforts to assess the impact of implementing smart metering, an up-to-date comprehensive analysis that considers both regulatory and quantitative aspects is still missing. This chapter contributes to fill this gap, undertaking an assessment of regulatory impacts and discussing methodologies to quantify costs and benefits. It provides information on the context of smart metering in Romania; defines the problems associated with the current framework; provides a list of possible aims and objectives in relation to smart-metering implementation; discusses a range of policy alternatives; provides information around the assessment of impacts in terms of costs and benefits. Hence, it allows assessing different options for the implementation, roll-out and calendar of smart metering in Romania.

Background: smart metering in Romania

For the past ten years, the Romanian government has attempted to assess the impacts of smart-metering implementation in Romania.

In 2012, it was estimated that implementing smart-metering systems in Romania 'had the potential to be a profitable investment', with a Net Present Value (NPV) of 1.17 billion RON and expected profitability of 44% over 2013–2032 (AT Kearney, 2012).

The Romanian energy regulator, Autoritatea Naţională de Reglementare în domeniul Energiei (ANRE), sought to gain knowledge of a broad range of issues and assumptions that were not known during the preparation of the AT Kearney (2012) study, particularly concerning technical, economic and administrative barriers in implementation; and optimal technical solutions and smart-metering functionalities, by areas and consumer types, to prioritize and ensure that implementation is technically feasible and financially sustainable and energy savings exceed costs. Before moving to an 80% roll-out, areas and types of consumers where smart meters would be implemented had to be prioritized based on network status, costs, economic and financial viability, before approving the full roll-out calendar.

Since 2012, following the implementation of the pilots and the delays in the original implementation schedule, several initial assumptions from the AT Kearney study have changed – particularly those regarding costs, measurable benefits and the actual calendar for the roll-out. As such, ANRE stated that they would reassess the options of an effective and efficient implementation plan for smart meters based on existing evidence.

Romanian law and regulation

The Romanian government made the implementation of smart metering mandatory through Law 123/2012 art. 66, provided that the cost–benefit analysis (CBA) shows positive impacts for the 'functioning of the energy market'. After the approval of the energy law in 2012, the regulator ANRE issued several Orders to implement pilot projects for smart metering and further refine the assumptions of the original study. Order 91/2013 defines smart metering and mandatory functionalities and started the implementation of pilot projects. Order 145/2014 replaces 91/2013 and brings some clarifications to ensure investments in smart meters are prudent; sets eligibility criteria for pilot projects proposed by Distribution System Operators (DSOs); defines CBA indicators for the pilots; and describes the monitoring of the implementation of smart-metering pilots. It was later amended through Orders 119/2015 and 6/2016.

Below is a summary of legislation and regulation which has been produced to data in relation to smart-metering implementation in Romania.

With regard to primary legislation, according to Law 123/2012, ANRE is mandated to evaluate potential for smart-metering implementation in Romania by September 2012 (i.e. at the time of the conclusion of the AT Kearney study) and prepare a roll-out plan to ensure an 80% roll-out by 2020, if the assessment concluded implementation is economically feasible. Implementing smart-metering systems is to be included in annual investment plans of the DSOs.

Law 123/2012 was amended several times. The current thinking is to have a roll-out by 2019–2024, but it remains unclear whether up to 80% or 100% coverage level. This would imply having a Fourth Regulatory Period for distribution, and it would allow for planning that minimises the risk of high spikes of tariffs. Under this scenario, ANRE is expected to develop appropriate regulation by 2018.

With regard to secondary legislation, Order 145/2014 defines the smart-metering system (meter, transformers, equipment to access the meter; information transmission systems; info management). It contains an annex with mandatory and optional functionalities; at the request of operators, functionalities can be relocated from mandatory to optional or optional to mandatory. The Order states the deadline of 80% roll-out by 2020 consistently with Law 123.

The Order with amendments regulates the pilot projects implementation. Pilots should not exceed 10% of the total annual investment plan. The Order contains data forms for pilot projects to be submitted for approval by ANRE and a schedule for monitoring data. In 2017 DSOs submitted plans for the roll-out to ANRE.

Costs of smart meters in Romania

Any assessment of the costs of smart metering in Romania is highly dependent on information about the market costs of the technologies involved, including the physical meters and ICT software and hardware. This section presents available data from the DSOs as part of the 2015–2016 pilot studies on smart-metering implementation. It discusses issues with smart-metering cost data. It explores possible solutions to the issues with smart-metering cost data, specifically with regard to capping smart-metering market price and depreciating smart meter costs.

Data from distribution system operators 2015–2016 pilot studies

Data on costs of smart meters derives from the pilot projects in 2015 and 2016. Specifically, non-sensitive commercial information is summarized in ANRE's Annual Report for 2016 published in June 2017.

Based on the initial Order for the pilots (145/2014), ANRE implemented smart meter pilot projects in 2015 and 2016, covering a variety of areas and customer types. The pilots covered residential and non-residential users connected to distribution grids; new, newly upgraded, as well as old networks; and rural and urban areas. Throughout 2015–2016, ANRE prepared five internal reports on the actual implementation of the smart-metering pilots, in an attempt to assess the viability of smart meters in Romania and to gauge the investment needs for the full (80%) roll-out, originally planned for 2017–2020.

To prepare for the roll-out, in February 2016 (Order 6/2016), ANRE set out to collect information on a full data set from pilots in both new and upgraded networks in urban and rural areas (2015 pilots) and non-upgraded networks in urban and rural areas (2016 pilots). Based on 2016 pilots in non-upgraded network areas, by 2017 ANRE should have had an assessment of the needs for investments in non-upgraded networks to integrate smart meters, to ensure that implementation of smart meters is feasible independent of the status of networks. The information collected, however, was not finalised in a CBA in 2017 (as previously planned). Instead, ANRE updated the AT Kearney study based on the pilot data. This was never realised in full because of major data gaps, particularly on benefits, and substantial differences on costs across operators resulting from the pilot data. By early 2017, ANRE had collected plans for roll-out from DSOs, with the intention to prepare the roll-out calendar in spring 2017, despite the missing data for the CBA. Pilots indicate that smart meters can be implemented also in areas with old and non-upgraded networks.

Figure 4.1 shows the variability of cost estimates from the pilots in terms of average unit cost per consumer and compares these with the initial assessment of costs in the AT Kearney (2012) CBA. Overall, the average cost per consumer varied significantly from the initial cost appraisal in the AT Kearney study. Actual costs in the 2015 pilots were 587 RON per consumer and 280 RON per consumer in the 2016 pilots. However, the three DSOs which have incurred the highest costs in the 2015 pilots (i.e. Transilvania Sud, 919 RON; Distributie Oltenia, 784 RON; and Muntenia Nord, 668 RON) chose not to participate at all in the 2016 pilots. Thus, the substantial reduction in total average costs of 52% is somewhat misleading. For DSOs participating in both 2015 and 2016 pilots, the cost reduction ranged from 16% to 49%, as shown in Table 4.3. Initial costs per consumer in the pilot projects submitted to ANRE for approval in 2016 were 434 RON. Initially, the AT Kearney study had estimated unit cost per consumer at 455 RON. The discrepancies are also due to differences across operators, as shown in Table 4.1.

Figure 4.2 shows a comparison of the breakdown of costs per meter between the findings of the pilot studies and the previous CBA. Overall, the

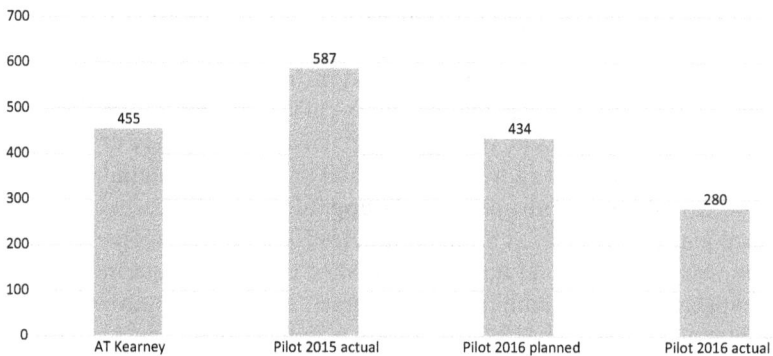

Figure 4.1 Average unit cost per consumer (RON).

Table 4.1 Summary of costs from pilots in 2015–2016

Distribution System operator	2015			2016			TOTAL		
	Pilot projects	No. clients	Pilot value	Pilot projects	No. clients	Pilot value	Pilot projects	No. clients	Pilot value
E-Distributie Banat	3	10,126	4,083,403	6	31,122	8,305,562	9	41,248	12,388,965
E-Distributie Dobrogea	4	10,227	3,928,854	4	26,565	7,936,769	8	36,792	11,865,623
E-Distributie Muntenia	1	11,016	3,940,472	4	50,539	13,215,654	5	61,555	17,156,126
Distributie Energie Oltenia	2	20,150	15,816,050	–	–	–	2	20,150	15,816,050
Delgaz Grid	2	22,622	7,913,352	2	48,721	14,265,570	4	71,343	22,178,922
FDEE Transilvania Sud	2	23,024	21,167,273	–	–	–	2	23,024	21,167,273
FDEE Transilvania Nord	2	5,470	3,232,573	2	8,210	2,480,500	4	13,680	5,713,073
FDEE Muntenia Nord	2	2,139	1,429,431	–	–	–	2	2,139	1,429,431
TOTAL	18	104,774	61,511,408	18	165,157	46,204,055	36	269,931	107,715,463

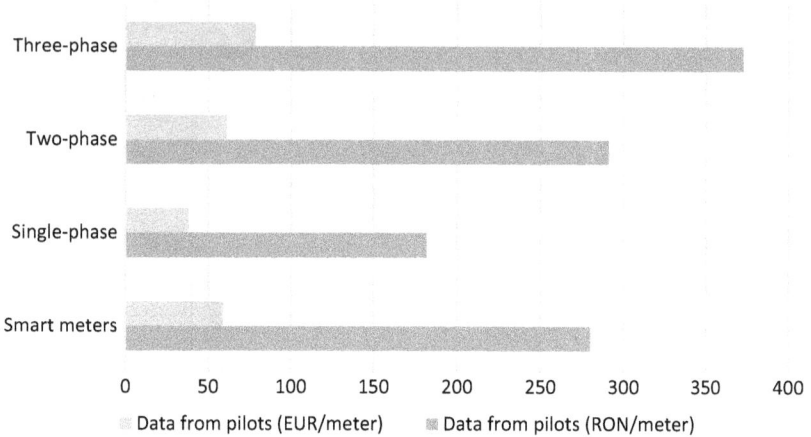

Figure 4.2 Break-down of costs per meter.

Table 4.2 Average cost by equipment type as of data from 2016 pilots

RON/meter	Single-phase meter	Three-phase meter	Balance meter	Communication cost	Investment cost
Gross average	182	291	373	44	280

Source: ANRE (2016). Raport Annual privind activitatea Autorității Naționale de Reglementare în Domeniul Energiei. http://www.anre.ro/ro/despre-anre/rapoarte-anuale

cost estimates of smart meter units decreased and so did the communications costs.

Table 4.1 shows the summary of costs from the 2015 to 2016 pilot studies. As mentioned above, CEZ (Distributie Energie Oltenia) and former state-owned Electrica subsidiaries (Transilvania Sud, Muntenia Nord) encountered difficulties in implementing pilot projects in 2015 and did not implement pilots in 2016. In 2015, their unit cost for smart meters was between 2 and 2.5 times higher than for the other operators. This might be caused either by differences in economies of scale and in prior experience with the implementation of smart metering or by problems with price disclosure in the tendering process (as further explained in Section 2.2).

Table 4.2 shows the average cost by equipment type as of data from all pilots up to the end of 2016. Balance meters represent the highest cost, followed by three-phase meters. Investment costs are calculated as total pilot value divided by the number of clients. For instance, the investment cost figure is derived as total pilot value (46,204,055 RON) divided by the number of clients (165,157 RON).

Table 4.3 Unit cost per meter: 2015 pilots and 2016 pilots

	Unit cost per consumer (RON/meter)		
Distribution system operator	2015	2016	%
E-Distributie Banat	403	267	34
E-Distributie Dobrogea	384	299	22
E-Distributie Muntenia	358	261	27
Delgaz Grid	350	293	16
FDEE Transilvania Nord	591	302	49
Total	587	280	52

Source: ANRE (2016). Raport Annual privind activitatea Autorității Naționale de Reglementare în Domeniul Energiei. http://www.anre.ro/ro/despre-anre/rapoarte-anuale

Table 4.3 compares the unit cost per consumer in the 2015 and 2016 pilots, taking 2015 as the base year. Costs decreased substantially between pilots in 2015 and 2016. It should be noted that three DSOs which did not implement projects in 2016 are not included.

Pilot projects in 2015 were implemented for an equivalent of 99.5% of planned physical meter installations and 88% of the planned value of installations. In 2016 pilots consisted of 97% of planned physical installations and 84% of the planned value of installations. The figure does not take into account the three DSOs which decided not to participate in the 2016 pilots. The lower percentage in value means that costs were below original estimates.

The pilots collected data for the following typologies of costs. The unit cost of the investment equals the investment total value, including the distribution network operations and electricity demand divided by the total number of end-customers. The unit cost of the investment needed for the single-phase meter purchase represents the investment for the single-phase meter purchase based on the number of single-phase meters installed with the customers of the project. The unit cost of the investment needed for the three-phase meter purchase equals the investment for the three-phase meter purchase based on the number of three-phase meters installed with the customers of the project. The unit cost of the investment for the system purchase and installation (without the meters) consists of the value of the investment for the system purchase divided by the total number of end-customers managed through the project. The unit cost of the investment for the purchase and installation of the information management meter subsystems is made of costs of servers, modems, database management system application and other auxiliaries. The unit cost of the investment for the purchase and installation of subsystems transmitting information is based on the costs of concentrators, signal repeaters and controllers. The unit cost of the investment for the works involving elements of the electricity network (without the meters) equals the investment value needed to ensure smart-metering operation, involving elements of the electricity distribution network divided by the total number of end-customers managed through the project.

The remainder of this section presents the main findings on costs of smart-metering installation of the 2015 pilots.

Figure 4.3 illustrates the unit investment costs for the implemented pilot projects. There are large differences between the minimum investment unit costs of 350 RON per customer and the maximum, which is 1,233 RON per customer. This means that the maximum investment unit cost is 250% higher than the minimum. The 18 pilot projects are grouped into five technical solutions (DSOs and equipment types) as follows:

– *ENEL*: ENEL Muntenia, ENEL Banat, ENEL Dobrogea, E.ON Distribuţie Romania
– *AEM*: CEZ Distribuţie, FDEE Transilvania Nord 2, Transilvania Sud 1
– *NES*: FDEE Muntenia Nord 2, FDEE Transilvania Nord 1
– *ELECTROMAGNETICA*: FDEE Muntenia Nord 1
– *ELSTER*: FDEE Transilvania Sud 2

Whilst some smart-metering equipment are common to several projects, unit costs vary significantly. A possible explanation is that the values of all DSOs were estimated based on the maximum bid values of the tender documents.

Since the outcome of the CBA of the AT Kearney (2012) study was positive for an investment unit cost of 99 Euro per customer, it results that only the pilot projects implemented by the ENEL and E.ON groups would yield positive net benefits under this cost.

The resulting average unit cost at national level amounts to 587 RON per customer, that is 30% higher than the limit unit cost in the AT Kearney

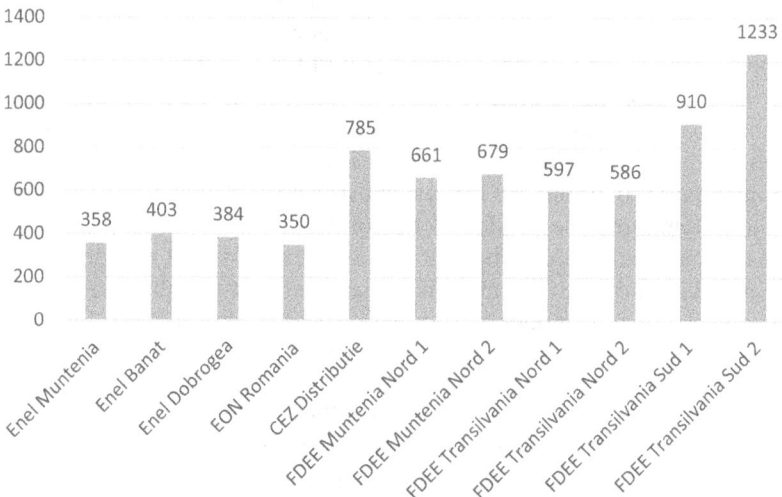

Figure 4.3 Unit investment cost per customer, 2015 pilots.

study, under circumstances in which about 70% of the customers are included in the "ENEL solution".

In accordance with recommendations provided by ANRE, the investment unit costs provided for in the pilot projects of CEZ Distribuţie and FDEE Transilvania Sud DSOs in 2016 were accepted by ANRE, with a maximum variation situated within the limits of ±20% of the average investment unit costs faced by DSOs related to smart meter implementation. These should not exceed 787 RON per customer. This is because the average unit cost amounts to 705 RON per meter. Consequently, for the pilot projects that do not comply not all investment costs were authorised by the regulator.

Figure 4.4 shows the investment of single-phase meters. These vary between 166 RON per meter and 753 RON per meter. A single-phase supply is smaller than other meters and cannot use as much power. A typical house will need a single-phase supply.

Figure 4.5 shows the investment costs for the purchase of three-phase smart meters, which vary from 277 RON per meter to 1947 RON per meter. Three-phase meters are larger and can use more power than other meters. For example, a larger house, flats or commercial building need three-phase meters.

Figure 4.6 shows the investment costs of balance meters. Balance meters combine distribution level information on energy demand and allow grid operators to better plan the integration of renewable energy into the grid and balance their networks. Hence, balance meters provide the possibility for consumers who produce their own energy to respond to prices and sell excess to the grid. Overall, the pilot studies point to very significant

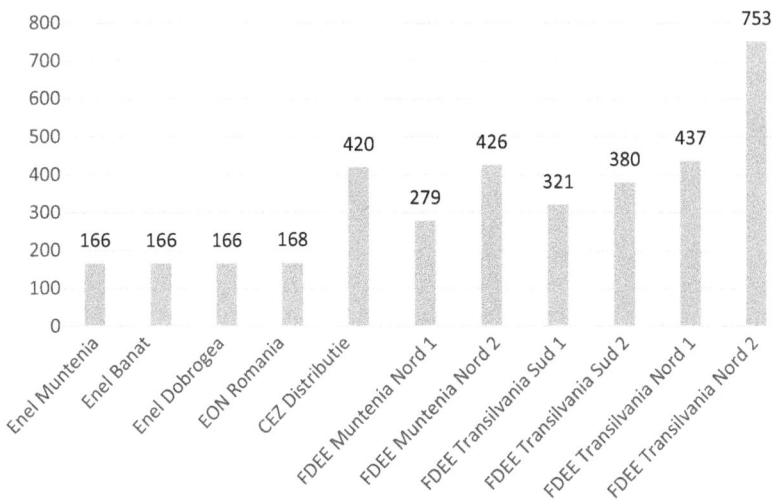

Figure 4.4 Investment costs for the purchase of single-phase smart meters in 2015 pilots.

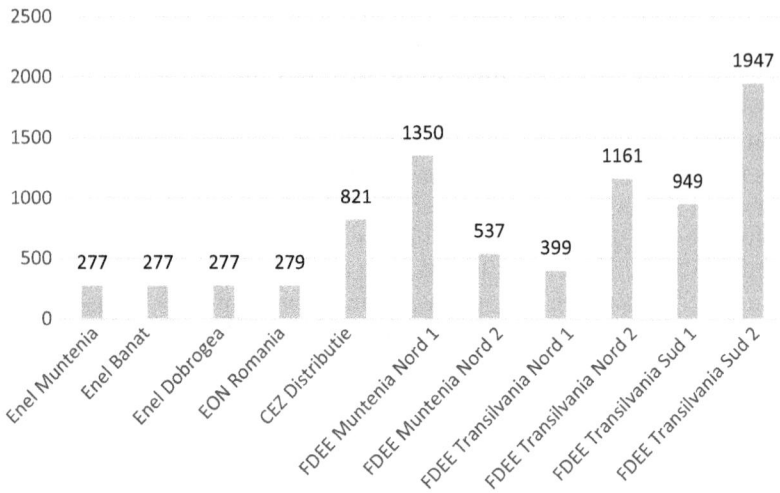

Figure 4.5 Investment cost for the purchase of three-phase smart meters.

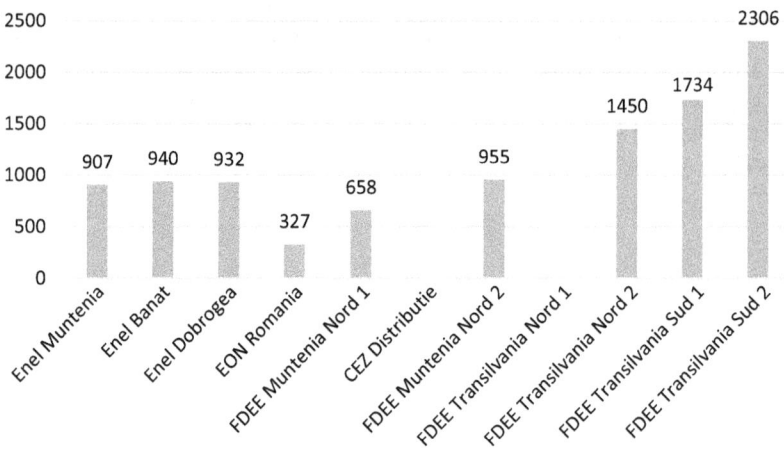

Figure 4.6 Investment cost for the purchase of balance meters, 2015 pilots.

differences between the investment costs for the purchase of meters. These costs do not include the labour costs. Substantial cost differences exist for the same supplier in different pilot projects.

Figure 4.7 illustrates the costs of the communication system per meter. There are great differences between the unit costs associated with different technical solutions. Power-line communication (PLC) communication was predominantly used for data transmission from the consumption location to the implemented smart meters in the 18 pilot projects. In three pilot

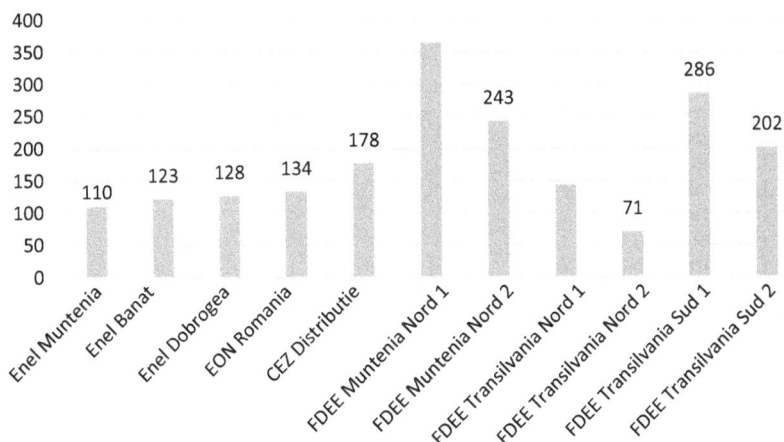

Figure 4.7 Cost of the communication system (per meter) in 2015 pilots.

projects (i.e. FDEE Transilvania Sud 1 – urban area; FDEE Transilvania Nord 2 – rural area; and FDEE Muntenia Nord 2 – rural area) GSM/GPRS and RF were also installed.

Out of the 18 pilot projects, the FDEE Muntenia Nord 1 project has a different technical solution as concentrators are not installed in the transforming stations, but in every block of flats, which increases their number and enhances significantly the investment cost.

In order to ensure the operation of the communication and transmission of the measured data, one of the implemented technical solutions required installing a significant number of auxiliary devices, signal repeaters and filters, which have increased the investment cost.

The situation regarding auxiliaries installed to ensure the communication operation is displayed in Table 4.4. The CEZ Distribuţie, FDEE Transilvania Nord 2 and FDEE Transilvania Sud 1 pilot projects have the same technical solution (i.e. supplier equipment).

Issues with smart-metering cost data

The data on costs for the 2015–2016 pilot studies on smart-metering implementation presents challenges which relate to the data collection, analysis and techniques for cost appraisal.

Two major challenges relate to capping smart meters' market price and metering depreciation costs. The pilots point to the fact that smart meter providers charge significantly different (250%) prices to different DSOs based on 'perfect foresight' of how much DSOs bided for in the tender documents. This poses questions as to how capping and competitive pricing could be in place for smart meters. There are also issues around the use

Table 4.4 Auxiliaries installed in the pilot projects in order to ensure communication

Distribution operator	Number of repeaters	Number of signal filters	Total
ENEL muntenia	9	0	0
ENEL banat	0	0	0
ENEL dobrogea	0	0	0
E.ON romania	0	0	0
CEZ distributie	542	815	1,357
FDEE muntenia Nord 1	0	0	0
FDEE muntenia Nord 2	0	0	0
FDEE transilvania Nord 1	0	0	0
FDEE transilvania Nord 2	62	167	229
FDEE transilvania sud 1	701	850	1.551
FDEE transilvania sud 2	0	0	0

Source: ANRE (2016). Raport Annual privind activitatea Autorității Naționale de Reglementare în Domeniul Energiei. http://www.anre.ro/ro/despre-anre/rapoarte-anuale

of depreciation and recognition in tariffs of the depreciation for the new meters and how this should be treated in a CBA on smart metering.

In addition to these two major challenges, the data on costs for the 2015–2016 pilot studies on smart-metering implementation features the following four issues.

First, the value of investment needed for balance meter is much more than the proposed/expected value in two instances. These seem to be driven by a higher volume (higher number of installations) and higher price per balance meter.

Second, the value of the investment needed for the purchase and installation of the information management subsystems and of the information transmission subsystems from the meters and the value of the investment needed for the purchase and installation of the information management subsystem from the meters do not generate consistent figures.

Third, the value of the investment needed for the purchase and installation of the information transmission subsystem are all much more than the proposed/expected value for one DSO. These seem to be driven by a higher price per communication modules and not by a higher volume (higher number of communication modules and auxiliary devices mounted in the system). Fourth, there are higher achievement rates for households than for non-domestic users.

In addition, the pilot projects conducted in 2015 did not integrate consumers who are also producers. This means that no data were collected on prosumers.

In order to understand how representative figures from trials are of the overall costs of roll-out in Romania, one should note that the 2015 and 2016 cover different networks (e.g. new and upgraded vis-à-vis old networks;

rural vis-à-vis urban networks; and households vis-à-vis non-households connected to distribution grids). This has a significant influence on the extent to which it can be inferred that by multiplying costs by full penetration rate the CBA will obtain credible costs of roll-out at national level. However, the spread of costs suggests that pilots were not all concentrated in lower implementation cost areas.

Benefits of smart meters in Romania

Data from distribution system operators 2015–2016 pilot studies

The most up-to-date data on benefits of smart meters derives from the pilot projects in 2015 and 2016. This is highlighted in ANRE's Annual Report for 2016 published in 2016.

The pilots collected data for the following typologies of benefits: cost reduction per customer for meter reading; cost reduction with interventions at the consumption locations; specific costs of the investments with the smart meter implementation; reduction of the commercial capacitive power transfer (CPT); and reduction of the technical CPT.

Table 4.5 lists smart-metering benefits from the findings of the 2015 pilot studies. As mentioned above, the 2015 pilots were carried out over sections of the network which were either new or recently upgraded. This means that any generalisation on benefits should take this into account. In addition, with regard to the reported data on benefits from the pilots, the following five observations should be taken into account. First, some of the information regarding the monitoring of the benefits is missing with several operators, which proves the lack of experience in managing such projects and a poor organization of the monitoring activity. Second, there is no consistency in the reported data: for instance, in two pilot projects there were recorded increases in the commercial CPT, contrary to the original estimations. Third, there was no correlation between the information prior to smart-metering implementation and the benefits obtained subsequent to smart-metering implementation. Fourth, there are significant value differences between operators, which leads to the conclusion that the assumptions or estimates made are based on different principles. Fifth, the estimation and calculation of the benefits obtained consequently to the implementation are an issue of the smart meter implementation process.

Additional typologies of benefits partly captured in the pilot studies consist of the reduction of the interruption duration in the electricity supply (to the consumer); the reduction the of the number of complaints on the measurement errors; the number of identifications of the excess of the contracted registered power by the system after having installed smart meter; the number of identifications of voltage variation beyond the accepted limits; variations in the monthly average consumption of electricity for household; consumers included in the smart meter implementation project; variations

Table 4.5 Smart meter main benefits (from 2015 pilot studies)

	Pilots	P1	P2	P3	P4	P5	P6	P7	P8	P9	P10	P11	P12	P13	P14	P15	P16	P17	P18
Reduction of costs for meter reading per customer RCcc	Costs with meter reading after smart meters	2.26	2.26	0.22	0.22	0.22	0.14	0.14	0.14	0.14	0.52	2.04	2.04	1	1.74	0.12	0.11	0	0
	Costs with meter reading before smart meters (RON/year)	9.02	15.28	7.25	7.25	7.25	7.65	7.65	7.65	7.65	10.38	4.56	4.56	4.77	7.08	1.19	1.77	0	0
		75%	85%	97%	97%	97%	98%	98%	98%	98%	95%	55%	55%	79%	75%	90%	94%		
Reduction of intervention costs at consumption points RCintc	Costs of intervention at consumption points after smart meters installation	2.26	2.26	0.09	0.09	0.09	0.08	0.08	0.08	0.08	0.33	8.32	8.32	0.14	0	0	0	0	0
	Costs of intervention at consumption points before smart meters installation	149.67	149.67	3.01	3.01	3.01	4.33	4.33	4.33	4.33	6.73	12.36	12.36	0.57	0	0	0	0	0
		98%	98%	97%	97%	97%	98%	98%	98%	98%	95%	33%	33%	75%					
Reduction of commercial losses RCPTcom	Commercial losses after implementation of smart meters	20.93%	15.35%	3.15	13.92	2.91	12.59	6.6	11.55	12.46	4.79	6%	11%	0.2	1.08	0	0.5	0	0
	Commercial losses before implementation of smart meters	15.52%	27.24%	3.72	16.92	2.45	13	9	12	13	5.45	17%	27%	1.33	5.4	2.93	1	0	0
		−35%	44%	15%	18%	−19%	3%	27%	4%	4%	12%	65%	59%	85%	80%	100%	50%		
Reduction of technical losses RCPTth	Technical losses after implementation of smart meters	12.40%	12.40%	8.28	9.09	7.96	7	5	7	9	10	13%	18%	9.53	7.6	7.02	0	0	0
	Technical losses before implementation of smart meters	15.70%	15.20%	8.28	9.09	7.96	7	5	7	9	10	15%	18%	10.99	14.3	7.02	7	0	0
		21%	18%	0%	0%	0%	0%	0%	0%	0%	2%	13%	0%	13%	47%	0%	0%		

in the monthly average consumption of electricity for non-household consumers included in the smart meter implementation project; variations in the consumption during peak hours for household consumers to whom smart meter was installed; and variations in the consumption during peak hours for non-household consumers to whom smart meter was installed.

Table 4.6 highlights the findings from the 2015 pilot studies on additional benefits of smart metering. There are no data and information for all the set indicators because, at this stage of the SM implementation, the distribution operators have not implemented the available IT applications of information processing or some projects have not been completed in 2016, the monitoring period being insignificant.

In summary, the pilot project present five main findings with regard to benefits. First, meter reading costs reduction varied between 55% and 97%. Second, the reduction of operating costs for activities requiring physical presence of specialised teams ranged between 27% and 97%. Third, CPT is associated with extreme variance, from an increase of 38% in CEZ – Distributie Oltenia to a reduction of 100% – total reduction of CPT in pilots from Transilvania Nord. It should be noted that CPT in AT Kearney (2012) was calculated based on formula and coefficients, not actual figures. Fourth, the reduction of complaints regarding metering errors varied between 4% and 100%. In one of the pilots for Transilvania Sud, there was an increase of complaints based on higher accuracy of metering.

Given the lack of data, ANRE considered the following benefits categories as non-measurable: preparation for the smart grids, flexibility of the networks, demand response, integration of prosumers and distributed generation, CPT reduction.

The Distribution System Operators, with ANRE's supervision, prepared a study on the status of the distribution, low-voltage networks. However, the findings of that study are rather general and basically indicate that without data from smart meters it is impossible to gather sufficient information with regard to the performance of distribution networks at the low voltage level. In other words, a decision on the implementation of smart meters would not be based on a CBA as most of the benefits (except the ones above identified during the pilots) cannot be currently estimated with more precision than at the time of the AT Kearney (2012) study.

Figure 4.8 illustrates the share of benefits as percentage of estimates associated with data from pilot studies. The largest benefits consist of reductions in technical losses, followed by reductions in O&M costs. Reduction in meter readings and operations, as well as reductions in outage times, presents limited benefits. The most significant missing benefit (also in comparison with the previous CBA) is reduction in communication losses. Indicatively, if adequately quantified and monetised, this benefit could be the second largest in the pie chart above and could push the total of benefits much closer to the cost figure. The benefits estimates relate to individual meters.

Table 4.6 Smart meter additional benefits (from 2015 pilot studies)

Pilots	P1	P2	P3	P4	P5	P6	P7	P8	P9	P10	P11	P12	P13	P14
Reduction of average interruptions in energy supply at consumer Redd_nealim														
Index average interruption in network (system) after implementation of smart meters	0	0	27.4	20.72	23.93	2.72	68.37	26.5	20.52	38.8	28	112	0	65
Index average interruption in network (system) before implementation of smart meters	0	0	35.57	26.59	37.97	51.88	230.25	15.91	83.43	53.47	27	104	0	69
			23%	22%	37%	95%	70%	-67%	75%	27%	-4%	-8%		6%
Reduction of complaints concerning metering errors Red_reder_mas														
Average yearly complaints concerning meter reading errors after implementation of smart meters	16	67	3	0	4	2	1	1	0	3	29	31	259	0
Average yearly complaints concerning meter reading errors before implementation of smart meters	23	73	2	9	16	3	2	2	0	51	58	56	187	2
	30%	8%	-50%	100%	75%	33%	50%	50%		94%	50%	45%	-39%	100%
Number of times that actual power exceeded contracted power after implementation of smart meters -Nr_deppcontr	n/a	n/a	10	9	58	304	236	371	39	178	492	450	0	0
Number of times of voltage variations beyond acceptable limits Nr.idvar_Un	n/a	n/a	0	0	0	0	0	0	0	0	597	572	0	0

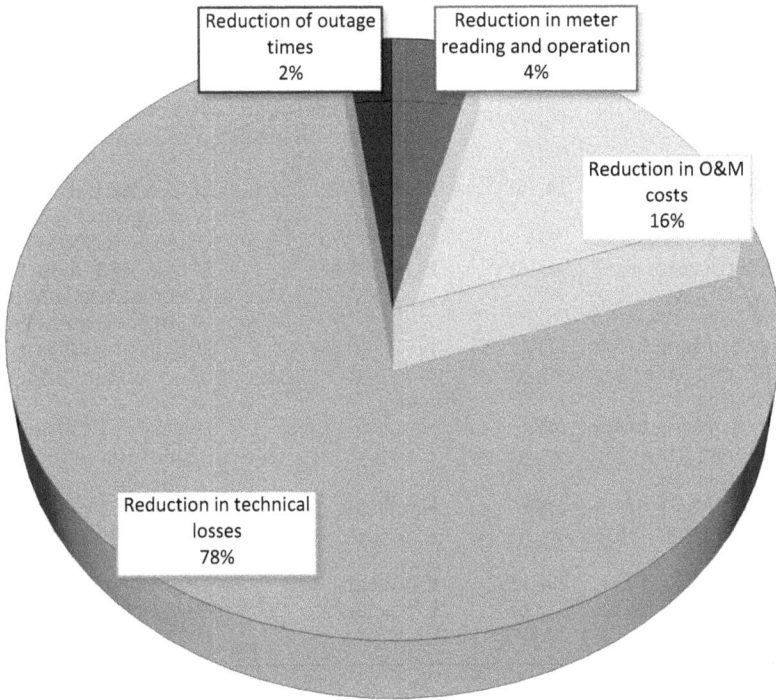

Figure 4.8 Share of benefits (data from pilot studies).

Table 4.7 shows the calculations for deriving reductions in technical losses. The methodology consists of obtaining figures for baseline CPT and average yearly electricity demand (per meter) from IEA statistics; impact of electrical losses from the 2015 ANRE report; the expected reduction in losses from smart-metering implementation; and expected reduction in losses from smart-metering implementation from the 2015 pilot studies. Finally current CPT, CPT reduction per meter and the value of CPT reduction were derived from my own calculations.

Figure 4.9 illustrates reductions in meter reading costs and O&M costs per meter based on data from 17 pilot studies. The average net reduction in meter reading costs was 7.15 RON per meter, whereas the net reduction in O&M costs was 26.94 RON per meter.

Identification of affected stakeholders

Identifying winners and losers from regulatory change is an important step in any CBA. Smart meter implementation affects the vast majority (if not the totality) of consumers, all actors in the electricity market and also manufacturers of smart-metering technologies. Overall, nine main groups of

Table 4.7 Calculations for deriving reductions in technical losses

Category	Value	Unit	Source
Baseline CPT	10.88%	Electric power transmission and distribution losses (% of output)	IEA Statistics © OECD/ IEA 2014 (iea.org/stats/index asp)
Impact of electrical losses	34.59	EUR/MWh	2015 ANRE report (Market of regulated contracts @ 140.56 RON/MWh)
Impact of electrical losses	0.03459	EUR/kWh	2015 ANRE report (Market of regulated contracts @ 140.56 RON/MWh)
Expected reduction in losses from smart-metering implementation	32%	%	2015 pilot studies
Expected reduction in losses from smart-metering implementation	26%	%	Average CPT from 2015 pilot studies with 80% roll-out
Average yearly electricity demand (per meter)	2584	KWh/cap	IEA Statistics © OECD/ IEA 2014 (iea.org/stats/index asp
Current CPT	281.1392	KWh/cap	Own calculations
CPT reduction per meter	823.65	KWh/cap	Own calculations
Value of CPT reduction	28.4900535	EUR/cap	Own calculations

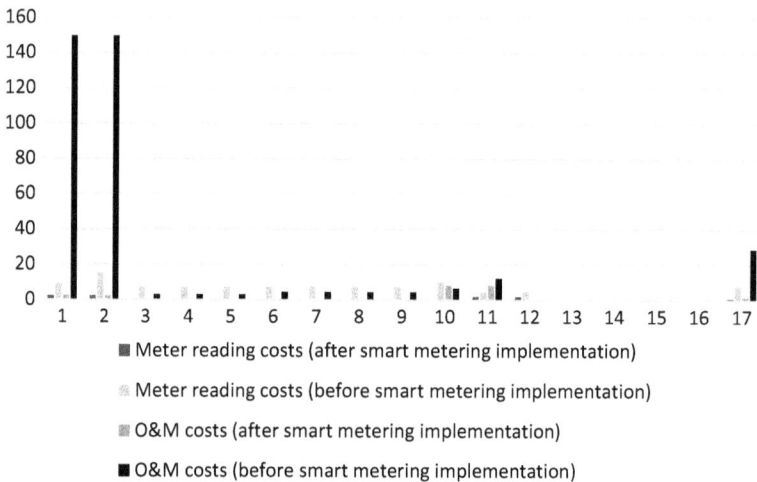

Figure 4.9 Reductions in meter reading costs and O&M costs (RON per meter).

stakeholders are listed below along with key metrics to assess and possibly quantify the impacts on each category of stakeholders. These were gathered as part of a Working Group on smart meters in Romania.

Implementing smart metering would allow DSOs to obtain better data on status of the grids. However, pacing is critical, to ensure that preparation, feasibility, design, procurement, contracting and actual installation follow the requirements of the new public procurement law and the labour and equipment markets have time to adapt to the new policy. Also, DSOs would have to optimise investment strategy given investment plans approved by ANRE and requirements for interoperability. Smart-metering implementation will also lead to better grid management, including safety (remote (dis)connection), and lower losses (CPT), but only in conjunction with other measures (commercial – policing; technical – other investments in the grid). Key metrics in this area include changes in manual meter reads, changes in costs of meter replacement, theft reduction.

In addition, one may consider "smart clients": demand response, optimization of consumption based on adequate information on consumption; better information to decide to switch suppliers; new technologies (e.g. electrical vehicles); better management of industrial electricity consumption, including aggregation of demand, demand response, etc. Key metrics for consumers consist of energy demand reduction, levels of prosumption through microgeneration.

Smart meters are essential to prepare the distribution infrastructure for the smart grids, better functioning of the balancing market and development of the market for renewables and energy produced by prosumers. ANRE will have to amend regulations and identify best means to recognize smart-metering costs in tariffs, e.g. setting caps (trade-off between tariff affordability and modernization of network); prepare regulations for distributed generation and small renewables; prepare regulations for electric vehicles and other consumers that would have more flexible access to electricity with the introduction of a better energy management at distribution level. All these regulations have to be properly correlated. They should be correlated also with the market liberalization and expected demand response gains. Key metrics for regulators include avoidance of multiple regulatory interventions (for instance, measured as number of 'one-in one-out').

Distributed generators, renewable providers and prosumers gain access to a new market once the distribution grids can take over the electricity they produce. A key metric for these types of stakeholders consists of the enablement of smart grid and greater use of renewables.

In principle, smart metering allows better forecasting, which means also optimization of purchases of electricity from the balancing market. This leads to better management of clients; optimized billing, disconnections of non-payers; optimization of energy procurement. Key metrics for energy retailers consist of more accurate billing, bad debt reduction, better forecasting.

Smart meter equipment manufacturers and communication/ICT companies see smart meters as a new market and may change business strategy once the regulatory approach on the implementation of smart meters becomes clearer. Key metrics for them include changes in turnover and changes in demand.

The main benefits for the government include broad societal gains on energy efficiency and energy market development (smart meters are part of the Climate Change Strategy and Energy Strategy currently under preparation). Higher employment (smart-metering installation) and higher tax revenues generated by smart-metering investments could also be considered. Key metrics for government departments consist of energy efficiency gains and CO_2 emission reductions.

The Romanian generation mix is well placed for balancing. Most of the energy it uses is derived from petroleum products (27.6%) and gas (27.1%), followed by solid fuels (17.9%), renewables (18.1%) and nuclear (9.1%). Some expensive (thermal – coal, gas) capacities are currently providing energy for balancing would no longer be in demand. Key metrics for generation companies include lower system marginal price as a result of lower demand and avoided costs of investments in new generation.

For the Transmission System Operator, the effects in terms of demand and supply balancing would become more visible with the introduction of smart meters. Key metrics for the Transmission System Operator are in terms of avoided costs associated with investigation of voltage complaints and avoided costs of re-enforcing the network as a result of reduction in peak demand.

References

ANRE Order 145/2014 (amended 4 times so far - Order 119/2015, Order 6/2016, Order 17/2017 and Order 31/2017).

Directive 2009/72/EC concerning common rules for the internal market in electricity and repealing directive 2003/54/EC.

Electricity and Natural Gas Law 123/2012 (Electricity and Gas Law -Art. 66, paragraphs 1 and 2).

Kearney, A. T. (2012). Smart metering in Romania: Final Report.

Poyry & Ciga Energy (2017). Evaluarea și monitorizarea retelelor de distributie din Romania - Raport pentru ACUE.

5 Cost–benefit analysis for energy policies

Introducing cost–benefit analysis

Over the past three decades, cost–benefit analysis (CBA) has been applied to various areas of public policies and projects, including energy. Research on CBA varies significantly and can be classified into two wide areas of work: studies which have attempted to define the technical-economic reasons underpinning CBA and studies which carried out empirical evaluations over the performance of samples of CBA.

From a theoretical point of view, CBA has been seen as a tool to increase the quality of regulation and public policy through welfare economics principles and Pareto efficiency. CBA in principle allows for the improvement of social and environmental conditions based on empirical evidence (Koopmans et al., 1964; Sunstein, 2002) whilst improving market competitiveness (Viscusi et al., 1987).

Empirical studies on CBA have focused on the choice of discount rate (Dasgupta, 2008; Gollier, 2002; Lind, 1995; Viscusi, 2007), the integration of distributional principles (Adler & Posner, 1999), the choice of datasets (Hahn & Litan, 2005; Morrall, 1986), the performance of different methodologies for monetising benefits and costs in cases where a market value does not exist (Sunstein, 2004; Viscusi, 1988). The latter point is of particular interest given the distance between theory and practice and deserves further reflection.

The impact of energy policy has been assessed with various appraisal and evaluation tools at least since the 1960s. Decision analysis, environmental impact assessment and strategic environmental assessment are all notable examples of progenitors of CBA in the assessment of energy policies, programmes and projects. This chapter provides an overview not only of CBA but also of other policy tools which have been historically applied to assess the impacts of energy policies, programmes and projects. It focuses on the types of data and models that typically inform CBAs for energy policies, the organisations involved and issues of data exchange between energy companies and policy-makers. Examples are derived from the European Commission, the United Kingdom, Italy, the Netherlands and France.

After this introduction, this chapter describes the historical development of CBA and classifies typologies of costs and benefits. CBA is the most comprehensive of a family of economic evaluation techniques that seek to monetise the costs and/or benefits of proposals. Following standard classifications, benefits and costs can be broadly defined as anything that increases human welfare (benefits) or anything that decreases human welfare (costs). The principles of efficiency are defined under CBA. The chapter will then move the focus to the specific application of CBA for energy policies. It will do so by describing available policy tools to assess energy policies, programmes and projects; identifying the institutions carrying out CBA on energy; examining issues of data quality in energy policies, before concluding.

Historical development of CBA

In order to understand the current scope and use of CBA in public policy-making (including in the energy sector), it is important to study how this tool has developed over time. CBA was originally designed as an interface between engineering and economics in areas of civil engineering which relate to energy policies and projects. More precisely, Dupuit, a French engineer and Marshall, a British economist, defined some of the formal concepts that are at the foundation of CBA. The Federal Navigation Act of 1936 required that the US Corps of Engineers should carry out projects for the improvement of the waterway system when the total benefits of a project exceeded the costs. This was initiated by Congress, which ordered agencies to appraise costs and benefits when assessing projects designed for flood control as part of the New Deal.

In the 1950s, economists tried to provide a rigorous, consistent set of methods for measuring benefits and costs and deciding whether a project is worthwhile. This mainly consisted in applying compensation tests and distributional weights. However, such measures were considered by several economists as a failure (Adler & Posner, 1999). Notwithstanding opposition, in the United States, CBA was increasingly applied in an expanding domain of policy areas, often following the rationale that alternative policy appraisal tools were less efficient (Pearce & Nash, 1981).

Following some experiences in Scandinavian countries and Canada, in the US Executive Order 12291 of 1981 institutionalised CBA as a consistent method for the appraisal of government policies and regulations, hence marking the beginning of the CBA era (Posner, 2000). To date there are soft-low requirements to conduct CBA on major policies and regulations in most OECD countries and examples from practice abound as highlighted below.

Assessing costs and benefits

The theoretical and practical implications of assessing costs and benefits in CBA practice are similar for energy and non-energy domains.

With regard to costs, each type of legislative change imposes various typologies of costs. Private companies, citizens and public administration can be subject to an increase in costs. The first significant classification is with regard to private and societal costs. The former consists of what a citizen or household has to pay in relation to a legislative change. CBA is often used by public administrations as an instrument to measure only certain components of private costs. This is particularly the case when legislative change is expected to have impacts on individual categories of companies.

Social costs represent what society as a whole has to pay in relation to legislative change. They typically include negative externalities and exclude transfer costs among groups of citizens (or companies).

Figure 5.1 outlines the typologies of costs associated with legislative change. Costs for public administration mean management costs as well as enforcement costs, i.e. costs associated with monitoring and inspections to ensure compliance. On the right side of Figure 5.1, private costs are divided between costs for private citizens and private companies. The latter is broken down in terms of direct financial costs, administrative costs, capital costs and efficiency costs.

With regard to benefits, the classification presented in Table 5.1 outlines some broad categories of benefits.

Economic benefits for which a value is provided in the market are not difficult to monetise. An example could come from a policy designed to add electricity generation from wind turbines. The market benefits of wind turbines are known because it is known both the physical amount of energy that the extra turbines would provide (i.e. kWh) and the monetary value of the physical quantity (i.e. €/kWh).

A contentious category of benefits consists of non-economic benefits which are not valued in the market that can be quantified and monetised. An example of this is reducing health risks. There is no market value for this and neither for saving lives, but the monetary value of the benefit can be seen as a reduction in the risk of dying or catching a disease. Economists

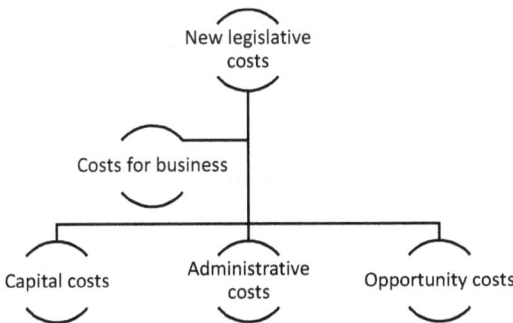

Figure 5.1 Typologies of legislative costs.

Table 5.1 Typologies of benefits

Types of benefits	Quantification and monetisation	Examples
Economic benefits valued in the market	Quantified, monetised	More wind turbines
Non-economic benefits which are not valued in the market that can be quantified and monetised	Quantified, monetised	Reducing health risks
Non-economic benefits that can be quantified but not monetised	Quantified, not monetised	Number of fish species saved from extinction
Non-economic benefits that cannot be quantified	Not quantified, monetised	More social justice

have developed four main methods of monetising non-market values associated with reductions in risk.

First, the 'willingness to pay values' techniques ask citizens how much they would pay to reduce the likelihood of a specific risk. In practice, this is implemented through stated preference surveys, where individuals are asked questions on changes in benefits; close-ended survey, where respondents are asked whether or not they would be willing to pay a particular amount for reducing risk; and stochastic payment cards, which offer to respondents a list of prices and associate likelihood matrix describing how likely the respondent would agree to pay the various offered prices.

Second, the 'human capital' approach calculates the value of a human life saved assessing the present value of the worker's earnings over the lifetime. The value is the benefit associated with reducing loss wages.

Third, the cost of illness (or medical costs assessment) method consists of an estimate of the costs to the medical system for treatment due to illness. Fourth, the 'willingness to accept values' techniques are based on the wage premiums workers are inclined to accept for any risks associated with their job. When the wage premium is divided by fatality risk, the result is the value of a statistical life saved.

Non-economic benefits that can be quantified but not monetised include, for instance, the number of fish species saved from extinction. CBA is centred on human lives, and impacts on other animal species are rarely taken into account in monetary terms as part of a policy appraisal, unless this refers specifically to ecological conservation and animal species protection. This is particularly the case for energy policies, which start from the assumption that energy is generated for and consumed by humans only. Nonetheless, handbooks on CBA, including the British HM Treasury's (2012) Green Book, may contain details about parameters to be used for plant species as part of the policy appraisal. The most recent version of the Green Book contains a detailed discussion on the use of natural capital as part of CBA practice.

Benefits which cannot be quantified and, for the same reasons, cannot be monetised include predominantly social benefits. An example consists of the benefit of improving social justice, thanks to a new policy or regulatory change. There might be social indicators which address some of this change, but this is hardly reconciled to a monetary value. In the US Executive Order 13563 on CBA, it is stated that agencies should take account of "human dignity" and "fairness," values, although these are "difficult or impossible to quantify".

Main concepts underpinning CBA

Understanding how CBA works and its fundamental principles is vital in order to comprehend its application to energy policy. The core efficiency principles of CBA lie in welfare economics (i.e. the branch of economic theory which has investigated the nature of the policy recommendations that the economist is entitled to make) within the domain of allocative efficiency (Baumol et al., 1977; Perman et al., 2003). Allocative efficiency (i.e. allocation of scarce resources that gives maximum social welfare) is defined via the concept of a 'Pareto improvement'. A Pareto improvement is a reallocation of resources (e.g. a decision to build a new power plant) that makes at least one individual better off without making anyone worse off. Pareto efficiency/ optimality is achieved when it is impossible to make one individual better off without worsening the condition of at least one other individual.

If economists restricted their domain of advice to Pareto improvements, they would not be able to advise on much, as most decisions involve a trade-off between making someone better off at the expense of making someone else worse off. What could be required is that the individual who gains must compensate the individual who loses for all of the latter's loss. If the individual who gains still gains after having paid out the compensation in whatever way (e.g. cash), the move would still be a Pareto improvement. This is not much less restrictive, because actual compensation is rarely paid. As an alternative, economists developed the idea of potential Pareto improvements. The Kaldor compensation test (after Nicholas Kaldor) sanctions a move from one allocation of resources to another, if the winner could compensate the loser and still be better off. In this case, the compensation does not actually have to be paid. Hicks identified a problem with Kaldor's compensation test – namely that it could sanction a move from one allocation to another, but it could equally sanction a move in the opposite direction, depending on where the problem starts. Instead, Hicks suggested that the loser could compensate the winner for forgoing the move, without being worse off than if the change took place.

A paper by Scitovsky (1951) unravelled the problem. Both rules need to be satisfied, such that a reallocation is desirable if, on the one hand, the winners could compensate the losers and still be better off and, on the other hand, the losers could not compensate the winners for the reallocation not occurring and still be as well off as they would have been if it did occur.

Because CBA is based on the Kaldor–Hicks efficiency criterion, it means that the benefits should be enough that those that benefit could in theory compensate those that loose out. It is justifiable for society as a whole to make some worse off if this means a greater gain for others. As stated above, under Pareto efficiency, an outcome is more efficient if at least one person is made better off and no one is made worse off. This is a stringent way to determine whether or not an outcome improves economic efficiency. However, some believe that in practice, it is almost impossible to take any social action, such as a change in economic policy, without making at least one person worse off (Buchanan, 1959). Using Kaldor–Hicks efficiency, an outcome is more efficient if those that are made better off could in theory compensate those that are made worse off, so that a Pareto improving outcome results.

In the case where the public sector supplies goods, CBA becomes a tool for judging efficiency (Stiglitz, 2000). The concept of efficiency, though, is normally thought on the premise of the market economy. This is particularly controversial when the decision involves some cost or benefit for which there is no market value or which, because of an externality, is not fully reflected in the market value.

Beyond CBA: other appraisal tools to assess energy policies, programmes and projects

CBA is a widespread tool in the domain of energy policy appraisal techniques, as discussed above. However, there are other ways to feed evidence into the formulation of energy policies, programmes and projects. This section provides an historical overview of alternative economic appraisal tools for energy policies, programmes and projects.

In the 1960s, decision analysis was first applied to investigate problems in oil and gas exploration. Its application was consequently extended from private to public sector (Huang et al., 1995). An early 1990s review enumerates 86 decision analysis studies that appeared in peer-reviewed journals from 1970 to 1989. Subsequent surveys found that decision analysis was frequently used to address strategic or policy decisions related to energy, such as investment options facing the utility industry, choice between different energy technologies, synthetic fuel policy, commercialisation of solar photovoltaic systems, management of nuclear waste and acid-rain control (Zhou et al., 2006). Decision analysis focuses on resolving conflicts between objectives, dealing with uncertainty about the outcomes and appraisal of multiple options. Given the fact that energy and environmental issues are generally complex and inevitably involve multiple objectives, the techniques involved in decision analysis varied depending on the level of uncertainty associated with the specific policy or project, with multi-criteria analysis cited more frequently.

Environmental issues became increasingly central to the development of energy policies and to the activities of the energy industry in the 1970s. This led to the upsurge of environmental impact assessment (EIA) and (later) strategic environmental assessment. The former was first a commonly accepted practice when developing energy infrastructure, and then became a regulatory requirement in several legal dominions around the world (Petts, 1999). The latter became an established practice in the late 1990s and over time improved its legal status in some jurisdictions (Dalal-Clayton & Sadler, 2005).

Environmental impacts of individual project proposals are the main focus of EIAs. Examples consist of new power plants or new hydroelectric dams (Mirumachi & Torriti, 2012). Legislation was first introduced in the United States in 1969. In Europe, EIAs were introduced, thanks to the EC Directive in 1985 (85/337), which was amended in 1997 (97/11). Currently, EIAs are carried out by institutions like the World Bank, the OECD member states, transition countries and several developing countries.

Because the environmental performance of the energy sector has been subject to higher degrees of scrutiny, questions were raised about whether EIA was the right tool to address the challenges associated with energy supply (Wood, 2003). Strategic environmental assessment (SEA) is designed to address environmental issues at a higher level of planning, which may take place at a regional, national and super-national scale. This is consistent with the idea that environmental protection needs to be embedded into energy frameworks at early phases of conception. The main origins of strategic environmental assessment in the EU relate to the Strategic Environmental Assessment Directive (2001/42). SEA is supposed to complement EIA for strategic actions. Strategic action is a more nebulous process than the formal submission of a development proposal, as in EIAs. Thus, SEAs address concepts rather than particular activities and must deal with incremental and non-linear policy processes (Wood & Dejeddour, 1992). Because it is focused on strategic actions, SEA is designed to include a stronger consideration of alternative options than EIA. The environment is often singled out in SEA, more so than in EIA or CBA, in large part because of the need to bolster its importance relative to the economic and social dimensions (Thérivel & Partidário, 1996). It is not clear which, if any, applications within the energy sectors require a strategic environmental assessment, and Jay (2010) notes that strategic environmental assessment has not been extensively adopted in the area of energy production.

This may be explained in relation to the fragmented nature of the industry, since generation, transmission, distribution and supply operate as separate markets – at least where liberalisation took place (Torriti, 2014). This makes the use of strategic planning tools more difficult. Today, strategic environmental assessment has potential in the fields of landscape, carbon reduction and air quality.

CBAs on energy: institutional differences

Government departments

Because of the importance of energy and its relationship with economic growth, energy policy sits firmly in any governmental agenda. The institutionalisation of energy policy often translates into the presence of energy ministries or energy departments within government (Newbery, 1989). These, like any other government departments, are charged with the task of formulating their policy with the support of CBAs. In addition, energy-related policies can be developed within departments for environment, industry and transport. To date no research has collected and let alone examined the body of CBAs produced by government departments. This section seeks to capture four salient features of CBAs by government departments. Four issues characterise the CBA on energy policies by government departments.

First, unlike CBAs by regulatory authorities, which feature a high level of techno-economic analysis (see section below), governmental CBAs on energy policy tend to follow a more generalist approach. CBAs conducted by government departments are often less quantitative in terms of the analysis and are geared to a less specialist audience. An example of this is the CBA on the Green Deal and Energy Company Obligation in the United Kingdom (DECC, 2011). The Green Deal aims to overcome access to capital and mismatched incentive problems. The Energy Company Obligation aims to provide additional support to deliver socially costs-effective measures that are not likely to be taken up under current policies, and provides measures to relieve fuel poverty. In essence, these are complementary policy measures intended to address barriers to the slow uptake of cost-effective energy efficiency measures. The government's CBA estimates that the Green Deal will lead to 125,000–250,000 households being lifted out of fuel poverty by 2023, but there have been criticisms with the way this figure was derived. More specifically, it has been argued that the CBA is too simplistic. For example, it neglects distributional issues: the Green Deal might increase fuel poverty since the policy might only benefit better-off end users and not be supportive of the fuel poor (Arie, 2012). The CBA was also criticised for applying very high discount rates (7%, which is significantly higher than other policies). Indeed, the CBA shows that investments do not generate positive Net Present Values for discount rates of 5%. However, other similar case studies show that the types of technological solutions contained in the Green Deal would create negative Net Present Values even with discount rates as low as 1.5%.

Second, there is a tendency to outsource research and analysis for those CBAs which require highly specialised electrical engineering and energy economics knowledge. For instance, in Ireland, Spain and the United Kingdom, consultants such as Frontier Economics, London Economics, Mott MacDonald, NERA and RedPoint have been contracted to carry out CBAs and to come out with policy options for CBAs on key policy areas such as smart metering, energy efficiency, renewable heat incentives, and feed in tariffs. The smart meter CBA by Mott MacDonald highlighted that the most advanced smart-metering options would have negative Net Present Values. It was noted that this initial negative assessment was partly due to assumptions that limit the value of demand-side management. Hence, the final CBA on smart meters presents a preferred roll-out option with a positive Net Present Value (DECC, 2009).

Third, the tendency to delegate pieces of analysis also results in interest groups gathering in specialist groups to produce the quantitative sections of a CBA. For instance, as part of the UK government's Electricity Market Reform, DECC asked a technical experts group, comprising the UK transmission system operator (National Grid), distribution network operators and energy aggregators, to produce analysis regarding the details of transitional arrangements to include demand-side response and energy storage within newly formed capacity mechanisms. Similarly, in the case of the UK policy for 'zero carbon homes', which is part of a more general approach to low energy buildings, leadership on CBA was given by the UK government to the Sustainable Building Task Group in 2003 (BIS, 2008). The group was co-chaired by the Chairman of the Environment Agency and the Deputy Chairman of English Partnerships and consisted of representatives from industry and environmental groups. According to Hauf (2012), a similar combination of government, industry and environmental groups united in the same policy formulation body also occurred in France. The specific proposals for the amendment of building regulations were developed by the Building Regulations Advisory Committee which produced the results of the accompanying CBA.

Fourth, the high political implications of energy policies mean that there are occasions where policies are pushed forward regardless of the 'better regulation' principles dictated by the same government. For instance, in 2011, the initiative by DECC of rolling out smart meters was the only example of policy escaping the one-in one-out rule applied by the UK coalition government. According to the one-in one-out rule, regulation whose direct incremental economic cost to business and civil society organisations exceeds its direct incremental economic benefit to business and civil society organisations can only come to place along with deregulatory measures whose direct incremental economic benefit to business and civil society organisations exceeds its direct incremental economic cost to business and civil society organisations (HM Treasury, 2011). In other words, the rule requires that no new national regulation is brought in without other regulation being removed. This also implies that the introduction of new regulations and removal of existing regulations are both government interventions and require their own separate CBA. The Smart Meters' initiative was classified as an "in" under the one-in, one-out methodology, because the CBA showed some £57m equivalent annual net cost to business. However, it was introduced as new legislation without any significant "outs". DECC stated that there was plan to simplify the nuclear decommissioning financing and fees framework, hence reducing paperwork burdens on operators (DECC, 2012).

Energy regulators

Since the 1990s in several developed and developing countries, the liberalisation of energy markets has been coupled with the emergence of energy regulatory agencies. CBA, along with stakeholder consultation, has become a common tool in liberalised markets and in markets which are in the process of being liberalised (Torriti & Ikpe, 2015). In some countries, the energy regulators stand out as a positive exception for having implemented CBA more rigorously than other government departments and other agencies (Renda, 2004). A review of CBA implementation across Italy confirms that the gas and electricity regulatory authority follows appropriate criteria (La Spina & Cavatorto, 2008).

Regulatory authorities have been driven by the dual aim of reducing prices to end users and improving the quality of energy supply. In order to obtain lower prices for end users, one of the main tasks of energy regulators is to regulate the prices of distribution companies, because these are considered regional monopolies and need incentives to ensure that they improve efficiency and raise the quality of supply. Energy regulators use price control reviews to regulate the prices that distribution companies can charge suppliers for transporting electricity through their networks (Cowell, 2004). The reviews normally take place every four to five years and involve a complex methodology which delivers data supporting the CBA.

The approach for the CBA commences with companies submitting a business plan setting out their operating costs, proposals for improving quality of supply and capital expenditure estimates for the next five years. The regulators typically enter these data into a series of cost benchmarking exercises, with companies' estimated expenditure benchmarked against each other. Given the importance of price control reviews in determining the development of distribution electricity systems, the accompanying CBA is arguably the most significant piece of regulatory analysis in any liberalised energy market (Ikpe & Torriti, 2018). For this reason, two issues are particularly worth of notice.

First, in spite of its highly technical features, the final CBA seldom contains much detail about the actual cost curves of distribution network operators. Issues of competition mean that regulators may not make explicit allowances for particular infrastructural projects. Thus, some companies may enter dialogue with the regulators during the review process, but find relatively limited justifications for the review decisions in the final CBA (Guy & Marvin, 1996).

Second, price review CBAs typically neglect non-techno economic impacts, including social and environmental impacts. For instance, Ofgem's (2001) Environmental Action Plan affirms that 'the choices made in the design of price control regulation can have wide ranging environmental impacts', but specifies what has long been the regulator's position on environmental and social matters: that it is not an environmental policy-maker and does not produce social policy.

Regulatory authorities cyclically conduct another example of major CBA on Quality of Supply regulation, which has similar rules in various European countries (e.g. United Kingdom, France and Italy). In Italy, this is subject to four-yearly revisions as part of which the regulatory authority sets the penalties and incentives for distribution network operators. The CBA process has been studied as an example of effective integration of various factors, including economic analysis based on end users' willingness to pay for better energy provision (Ajodhia et al., 2006) and consultation (Fumagalli & Lo Schiavo, 2009). Torriti et al. (2009) describe the CBA process for two reviews of the Quality of Supply regulation in terms of preliminary analyses, research studies, alternative regulatory options, consultation and CBA. They highlight some of the analytical and procedural issues typically associated with CBA: the creation of alternative options, the development of CBA, the disparity between analytical effort and available resources and the need to communicate in an informed manner with interested stakeholders. The experience from this example also shows that when attention is paid to these details, CBA can generate unexpected results.

European commission

A significant share of policies produced by the European Commission is in the area of energy, several of which require an Impact Assessment and, consequently, a CBA.

Before the creation of DG Energy in 2010, its predecessor – DG for Transport and Energy (TREN) – conducted most of the CBAs on energy. However, over the years, CBAs on energy policies have been carried out also by DG Environment, Climate, Informatics (DIGIT) and Economic and Financial Affairs (ECFIN). After the year 2009, the intensification of policy-making activity around climate change targets, with renewable energy, energy efficiency and lighting policies, justify the higher number of CBAs.

The level of analysis varied over time. What Hanley and Spash (1993) stated at the beginning of the 1990s that in the area of energy and environment benefits have not always been well integrated into the European Community policy assessments cannot apply to present times. Individual CBAs received some attention by researchers with regard to their economic assessments. For instance, the CBA on the EU's Objectives on Climate Change and Renewable Energy for 2020 (European Commission, 2008) considers the actual EU partitioning between Emission Trading Scheme (ETS) and non-ETS sectors to be cost-effective, whereas the CBA did not take into account the excess costs associated with differential emission pricing (Böhringer et al., 2009; Torriti, 2010).

Quality of data

One of the main issues around CBAs on energy regards the extent to which data from energy companies are used as part of the appraisals. Most of the energy regulators monitor the markets in order to foresee critical issues, prevent disturbances and enact timely regulatory actions. To ensure this task, they need significant amounts of data from market actors. Ideally, CBAs on energy could contain all the available information on markets performance; compare operations over time and across markets; publicly release all data submitted to and produced by the market and system operators; and even anticipate instances where small market flaws may develop into market failures. However, there are some obstacles to the transparent exchange of information between energy companies and energy regulators (Torriti, 2012).

There are data which companies have a right to maintain confidential. For instance, with smart meter data from the Energy Demand Response Programme (EDRP) in the United Kingdom, the only publicly available version of the dataset is void of sociodemographic information. The EDRP documents the results of trials conducted in the United Kingdom by four utilities in 2007–2010. The trials entailed a number of "treatments", including TOU tariffs, presentation of usage information in various ways, in-home displays, etc. Control groups were also included. The public-use version of the EDRP dataset does not contain any information about the trial a household was assigned to or the tariffs if the trial involved pricing. It simply documents kWh of electricity and gas used at half-hourly intervals. Minimal geographical information is available, along with the ACORN classification of the household and the fuels used. There is also one dummy variable stating whether the household is on a smart meter for gas, and another keeping track of whether it is on a smart meter for electricity. The removal of detailed information about the dwelling (detached, semi-detached, terraced home, etc.; internal floor space; construction materials; presence of insulation, details about the heating system, including age of the heating system, etc.), the household (number of household members, ages, gender and education, worker or otherwise status, etc.), supply status prior to the trial (credit, monthly debit, prepay, economy 7, etc.) and the trial itself (control or treatment status of the household, type of trial, pricing structure, etc.) makes it difficult for researchers to make use of this type of publicly available smart-metering dataset.

There are also data which are public in principle but costly to gather and assemble. There is no general consensus on the desirability of data exchange. Ofgem's guidance on CBA admits the challenges of capturing competition effects of regulatory change.

> In the case of markets being opened up to competition, for example, it is
> inherently difficult to predict with any accuracy the potential efficiency

benefits that introducing a competitive process might bring, or to quantify meaningfully the dynamic benefits of competition such as the scope for increased innovation and the introduction of new products, services and technologies.

(Ofgem, 2013, p. 23)

In principle, the exchange of information from energy companies to policy-makers is desirable because lack of exchange of data leads to incomplete information and inefficient screening of the market (Brown, 2001). According to this view, under publicly available CBAs, the information these disclose should be available not only for those who have the legal and financial capability to access data, but also to all market and non-market actors. However, in practice, incumbents argue that there are instances where transparency may violate property rights or harm business when the disclosure of crucial information can alter the competitive process (Campbell & Lindberg, 1990). An example of this relates to the treatment of electricity consumption data from smart meters. In principle, if regulators had access to such data, they could make informed decisions about tariffs, based on actual end users' consumption. In reality, the consultation and CBAs conducted in the Netherlands show that both energy companies and consumers were opposed to the disclosure of consumption data (Cavoukian et al, 2010). Moreover, even new entrants to the market may find data exchange problematic. Perfect visibility of the strategies of competitors may be beneficial to the defence of market power by the dominant company. Some delay in making market bids transparent may help the strategies of new competitors. In the Electricity Market Reform in the United Kingdom, DECC proposed to publish historical data on the bidding prices for the Short Term Operating Reserve, a service for the provision of additional active power from generation and/or demand reduction, and energy aggregators, which have only entered energy markets in the past five years, opposed such change.

Compared with the European Commission's DG Energy, the US Federal Energy Regulatory Commission features a higher legislative power for access to data and a greater financial capacity to purchase data. Since data are kept confidential by the regulator, business concerns of being negatively affected by data disclosure for competition purposes are limited. This arrangement implies that in the United States the transparency of CBAs is sacrificed in support of effective market monitoring and higher quality of data and analysis.

In the EU, institutional market monitoring activities are lagging. In some European countries, the regulator is recipient of a wealth of data (Gilbert et al., 2002), and the main issue is whether to publish them in a CBA or not. In other countries, data gathering does not represent a problem, but organisational deficiencies make the treatment of data rather difficult (Schweber et al., 2015).

Conclusion

This chapter described the context, theories and main features of and alternatives to CBA when applied to energy policies. The discussion on the historical development of CBA shows that this appraisal tool represents a mix of methodologies from engineering and economics which is suitable, at least in principle, to most energy policy problems. CBA is most directly applicable to civil engineering projects and programmes which rely on quantifiable units of cost. The chapter provides a breakdown of typologies of costs and benefits, hence noting some of the difficulties associated with the practice related to benefits appraisal in energy problems. Most economic appraisal techniques tend to treat net benefits (or undiscounted cash flows) as given. However, gathering data on future flows of benefits is an intrinsically uncertain exercise. CBAs on energy policies have traditionally relied on different modes of data collection and statistical inference, based among other things on the type of institution conducting the economic appraisal (e.g. central government departments, European Commission or regulatory authorities).

This is one of the reasons why the chapter addresses what economic efficiency means for CBAs associated with energy policies. This chapter reviewed economic efficiency as the core principle underpinning CBA, as it is a method by which this concept of efficiency can be applied to publicly supplied goods. Assuming that energy policies and projects bring utility to people, their Willingness to Pay represents the benefit of supplying such goods. In energy policy appraisals difficulty exists, however, in transferring the efficiency concept for market goods to publicly supplied goods. Two relevant points involve the efficiency concept in CBA. Energy policies may be concerned with a much broader range of consequences than those related to firms. Also, energy suppliers may not always use market prices in evaluating projects either because the market prices may not exist or because market prices may not represent true marginal social benefits/costs. The efficiency criteria on which CBAs for energy policies are based are relatively different from the traditional Willingness to Pay and Willingness to Accept relation because they are supposed to take into account distribution and equity issues. When efficiency conflicts with other values, it is actually impossible to create economic welfare criteria that integrate all values. An example of this discussion on the impractical use of CBA in the context of the macroeconomics of energy policy comes from projects aimed to address fuel poverty. These create both winners and losers. In most of the cases, losers already belong to the poorest and most marginalised members of society. Whilst fuel poverty projects can bring enormous benefits to society, their costs to the poorest have effects on their health, and even their lives (Kanbur, 2002). The use of CBA by government departments for fuel poverty projects is widespread and often criticised for not considering areas like basic needs approaches, shadow prices, social discount rates and macroeconomic shocks to public goods (Brent, 1998; Kirkpatrick & Weiss, 1996).

The examples given in this chapter offer a picture of the range of CBAs applied to energy policies. This chapter has observed how the level of techno-economic analysis varies in CBAs conducted by regulators, government departments and the European Commission. However, no judgement is placed here on the value of less technical CBAs. Indeed, a lower level of analysis may yield positive benefits in terms of greater engagement with stakeholders and the wider public in the development of policies that will have significant consequences for society as a whole. This benefit is even greater when taking into account the detachment of the lay public from energy policies (Cotton & Devine-Wright, 2012). Unlike some of the more technically focused exercises that have been used to assess energy regulation, CBA for energy policy is intended to be an inclusive and participative process, in which there is an opportunity for deliberation and consensus-building. It has been pointed out elsewhere that the application of CBA to energy policies emphasises three typical weaknesses consisting of the exclusive concern with economic values, the treatment of uncertainty, and the neglect of intergenerational effects (Simpson & Walker, 1987).

Very much like environmental policies, energy policies tend to be designed to achieve multiple objectives ranging from climate change to utilities' tariffs. Correspondingly, the impacts generated by energy policies tend to vary substantially in nature and size. Over the years, policy-makers have expanded the types of analytical tools used in the appraisal and evaluation of energy policies from narrowly scoped geophysical and ecological models, on the one hand, and purely socio-economic oriented tools of decision analysis, on the other hand, to highly integrated assessment tools, such as CBA. Through an expansion of geographic and temporal scopes and depiction of large complex systems, CBA models were expected to overcome many of the short-comings of earlier analyses, i.e. absent or inadequate depiction of technological change, micro-behaviours of economic actors (e.g., firms and consumers), intergenerational trade-offs and fairness and uncertainty (Greening & Bernow, 2004). However, given their broad approach, CBAs are prone to the problem of finding a balance between quantification of current economic and physical phenomena, and future variations in the supply and demand of energy systems.

References

Adler, M., & Posner, E. (1999). Rethinking cost-benefit analysis. *Yale Law Journal, 109*, 165–247.

Ajodhia, V., Lo Schiavo, L., & Malaman, R. (2006). Quality regulation of electricity distribution in Italy: An evaluation study. *Energy Policy, 34*, 1478–1486.

Arie, S. (2012). Understanding the risks of the green deal. Working Paper, Smith School of Enterprise and the Environment, University of Oxford.

Baumol, W. J., Bailey, E. E., & Willig, R. D. (1977). Weak invisible hand theorems on the sustainability of multiproduct natural monopoly. *The American Economic Review, 67*(3), 350–365.

BIS. (2008). Sustainable buildings task group. Retrieved from http://webarchive.nationalarchives.gov.uk/+/http://www.berr.gov.uk/sectors/construction/sustainability/sbtg/page11919.html

Böhringer, C., Löschel, A., Moslener, U., & Rutherford, T. (2009). EU climate policy up to 2020: An economic impact assessment. *Energy Economics, 31*, 295–S305.

Brent, R. J. (1998). *Cost-benefit analysis for developing countries*. Cheltenham: Edward Elgar Publishing Ltd.

Brown, M. A. (2001). Market failures and barriers as a basis for clean energy policies. *Energy policy, 29*(14), 1197–1207.

Buchanan, J. M. (1959). Positive economics, welfare economics, and political economy. *Journal of Law and Economics, 2*, 124–138.

Campbell, J. L., & Lindberg, L. N. (1990). Property rights and the organization of economic activity by the state. *American Sociological Review, 55*(5), 634–647.

Cavoukian, A., Polonetsky, J., & Wolf, C. (2010). Smart privacy for the smart grid: Embedding privacy into the design of electricity conservation. *Identity in the Information Society, 3*(2), 275–294.

Cotton, M., & Devine-Wright, P. (2012). Making electricity networks "visible": Industry actor representations of "publics" and public engagement in infrastructure planning. *Public understanding of science, 21*(1), 17–35.

Cowell, R. (2004). Market regulation and planning action: Burying the impact of electricity networks? *Planning Theory and Practice, 5*(3), 307–325.

Dalal-Clayton, B., & Sadler, B. (2005). *Strategic environmental assessment: A sourcebook and reference guide to international experience*. London: Earthscan.

Dasgupta, P. (2008). Discounting climate change. *Journal of Risk and Uncertainty, 37*, 141–169.

DECC. (2009). Impact assessment of a GB-wide smart meter roll out for the domestic sector. Retrieved from www.decc.gov.uk/assets/decc/consultations/smart%20metering%20for%20electricity%20and%20gas/120090508152831e@@smartmeteCBAdomestic.pdf

DECC. (2011). The green deal and energy company obligation impact assessment. Retrieved from https://www.gov.uk/government/uploads/system/uploads/attachment_data/file/43000/3603-green-deal-eco-ia.pdf

DECC. (2012). DECC's 4th statement of new regulation. Retrieved from https://www.gov.uk/government/uploads/system/uploads/attachment_data/file/48461/5787-decc-4th-statement-of-new-regulation.pdf

European Commission. (2008). Impact assessment, document accompanying the package of implementation measures for the EU's objectives on climate change and renewable energy for 2020. SEC(2008) 85/3, (23 January 2008). Brussels: European Commission.

Fumagalli, E., & Lo Schiavo, L. (2009). Regulating and improving the quality of electricity supply: The case of Italy. *European Review of Energy Markets, 3*(2), 1–27.

Gilbert, R., Neuhoff, K., & Newbery, D. (2002). Mediating market power in electricity networks. Working Paper, Department of Economics, University of Berkeley.

Gollier, C. (2002). Discounting an uncertain future. *Journal of Public Economics, 85*, 149–166.

Greening, L. A., & Bernow, S. (2004). Design of coordinated energy and environmental policies: Use of multi-crite CBA decision-making. *Energy Policy, 32*(6), 721–735.

Guy, S., & Marvin, S. (1996). Disconnected policy: The shaping of local energy management. *Environment and Planning, 14*, 145–158.

Hahn, R., & Litan, R. (2005). Counting regulatory benefits and costs: Lessons for the US and Europe. *Journal of International Economic Law, 8*(2), 473–508.

Hanley, N., & Spash, C. L. (1993). *Cost-benefit analysis and the environment.* Cheltenham: Edward Elgar.

Hauf, H. (2012). A collaborative approach to policy: A case study on the Zero Carbon Hub: 2020 Public Services Hub at the RSA.

HM Treasury. (2011). One-in, One-out (OIOO) Methodology. HM Treasury. Retrieved from https://www.gov.uk/government/uploads/system/uploads/attachment_data/file/31616/11-671-one-in-one-out-methodology.pdf

HM Treasury. (2012). *Accounting for environmental impacts: Supplementary green book guidance.* London: HM Treasury.

Huang, J. P., Poh, K. L., & Ang, B. W. (1995). Decision analysis in energy and environmental modeling. *Energy, 20*(9), 843–855.

Ikpe, E., & Torriti, J. (2018). A means to an industrialisation end? Demand side management in Nigeria. *Energy Policy, 115*, 207–215.

Jay, S. (2010). Strategic environmental assessment for energy production. *Energy Policy, 38*, 3489–3497.

Kaldor, N. (1939). Speculation and economic stability. *The Review of Economic Studies, 7*(1), 1–27.

Kanbur, R. (2002). Economics, social science and development. *World Development, 30*(3), 477–486.

Kirkpatrick, C. H., & Weiss, J. (eds.). (1996). *Cost-benefit analysis and project appraisal in developing countries.* Cheltenham: Edward Elgar Publishing.

Koopmans, T. C., Diamond, P. A., & Williamson, R. E. (1964). Stationary utility and time perspective. *Econometrica, 32*, 82–100.

La Spina, A., & Cavatorto, S. (2008). *Le autorità indipendenti* (Vol. 573). Bologna: Il Mulino.

Lind, R. (1995). Intergenerational equity, discounting, and the role of cost-benefit analysis in evaluating global climate policy. *Energy Policy, 23*, 379–389.

Mirumachi, N., & Torriti, J. (2012). The use of public participation and economic appraisal for public involvement in large-scale hydropower projects: Case study of the Nam Theun 2 Hydropower Project. *Energy Policy, 47*, 125–132.

Morrall, J. (1986). A Review of the Record. *Regulation, 2*, 25–34.

Newbery, D. (1989). Energy policy issues after privatization. In D. Helm, J. Kay, & D. Thompson (eds.), *The market for energy* (pp. 30–54). Oxford: Clarendon Press.

Ofgem. (2001). *Environmental action plan.* London: Ofgem.

Ofgem. (2013). Guidance on Ofgem's approach to conducting impact assessments. Proposed guidance. Retrieved from https://www.ofgem.gov.uk/ofgem-publications/82590/proposedguidanceonofgemsapproachtoconductingimpactassessments.pdf

Pearce, D. W., & Nash, C. A. (1981). *The social appraisal of projects: A text in cost-benefit analysis.* London: Macmillan.

Perman, R. (Ed.). (2003). *Natural resource and environmental economics.* London: Pearson Education.

Petts, J. (Ed.) (1999). *Handbook of environmental impact assessment.* Oxford: Blackwell.

Posner, R. A. (2000). Cost-benefit analysis: Definition, justification, and comment on conference papers. *The Journal of Legal Studies, 29*(S2), 1153–1177.

Renda, A. (2004). Qualcosa di nuovo nell'AIR? Riflessioni a margine del dibattito internazionale sulla better regulation. *L' industria, 26*, 331–364.

Schweber, L., Lees, T., & Torriti, J. (2015). Framing evidence: Policy design for the zero carbon home. *Building Research and Information, 43*(4), 420–434.

Scitovsky, T. (1951). The state of welfare economics. *The American Economic Review, 41*, 303–315.

Simpson, D., & Walker, J. (1987). Extending cost-benefit analysis for energy investment choices. *Energy Policy, 15*(3), 217–227.

Stiglitz, J. E. (2000). Capital market liberalization, economic growth, and instability. *World development, 28*(6), 1075–1086.

Sunstein, C. (2002). *The Cost-benefit state: The future of regulatory protection.* Chicago: American Bar Association.

Thérivel, R., & Partidário, M. R. (1996). *The practice of strategic environmental assessment.* Abingdon: Earthscan Publications Ltd.

Torriti, J. (2010). Impact Assessment and the liberalisation of the EU energy markets: Evidence based policy-making or policy based evidence-making? *Journal of Common Market Studies, 48*(4), 1065–1081.

Torriti, J. (2012). Multiple-project discount rates for cost-benefit analysis in construction projects: A formal risk model for microgeneration renewable energy technologies. *Construction Management and Economics, 30*(9), 739–747.

Torriti, J. (2014). Privatisation and cross-border electricity trade: From internal market to European Supergrid? *Energy, 77*, 635–640.

Torriti, J., Fumagalli, E., & Lo Schiavo, L. (2009). L'AIR nella pratica di una Autorità indipendente: l'esperienza dell'Autorità per l'energia elettrica e il gas di applicazione dell'AIR alla regolazione della qualità del servizio. *Mercato Concorrenza Regole, 10*(2), 285–321.

Torriti, J., & Ikpe, E. (2015). Administrative costs of regulation and foreign direct investment: Evidence from standard cost model implementation in non-OECD countries. *Review of World Economics, 151*(1), 127–144.Viscusi, W. K. (1988). Irreversible environmental investments with uncertain benefit levels. *Journal of Environmental Economics and Management, 15*, 147–157.

Viscusi, W. K. (2007). Rational discounting for regulatory analysis. *University of Chicago Law Review, 74*, 209–246.

Viscusi, W. K., Magat, W., & Huber, J. (1987). An investigation of the rationality of consumer valuations of multiple health risks. *RAND Journal of Economics, 18*, 465.

Wood, C. (2003). *Environmental impact assessment: A comparative review.* London: Pearson Education.

Wood, C., & Dejeddour, M. (1992). Strategic environmental assessment: EA of policies, plans and programmes. *Impact Assessment, 10*(1), 3–22.

Zhou, P., Ang, B. W., & Poh, K. L. (2006). Decision analysis in energy and environmental modeling: An update. *Energy, 31*(14), 2604–2622.

6 Meta-analysis of smart-metering trials and conservation effects

Smart meters and in in-home displays: do they reduce energy demand?

Smart devices are often paired with in-home displays, which provide near real-time feedback on aggregate home electricity consumption in terms of kWh and cost and require clip-on current sensors installed into the breaker panel of the home. Large-scale trials from 2000 onwards used devices which do not require installation in the breaker panel, and instead uses an optical sensor attached directly to the meter.

In-home displays communicate directly with smart meters, often using an open standards-based communication protocols, which allow for feedback and eliminate the requirement for costly or time-consuming installation processes. In-home displays also make use of graphical user interfaces.

These are the technologies which are supposed to deliver direct feedback to end users, but what is direct feedback? The next sections look at principles and empirical evidence from previous reviews in relation to this critical question.

In principle

According to Darby (2006, p. 3) direct feedback is defined as "immediate, from the meter or an associated display monitor". Early studies on direct, continuous feedback involved in-home displays (e.g. Energy Cost Indicator, or The Energy Detective in Hutton et al., 1986; Parker et al., 2006).

Behavioural scientists, especially in psychology and economics, have been debating on for some time is how people respond, in terms of energy demand, to smart meters. A set of demographic and attitudinal variables may influence conservation behaviour and the level of reduction from real-time electricity consumption feedback, though there is mixed evidence as to how much and in what direction. It has been argued that higher income is correlated to higher electricity consumption and higher environmental awareness. This, however, may not bring about more conservation from smart meters (Brandon & Lewis, 1999). It may be attributed to the fact that high income groups are more likely to engage in one-time, capital cost intensive home improvements that reduce home energy efficiency (Cunningham & Joseph, 1978). Conversely, a 1993 feedback study found no conservation effect in flats and low-income areas (Nielsen, 1993).

Smart meters may have a greater effect on energy consciousness (i.e. sticking labels on at the supermarket) rather than triggering actual energy demand reductions. Consistently, higher levels of environmental awareness have generally not been conducive of more conservation (Torriti et al., 2010). This may be because in empirical studies, the households that sign up or opt into the program are more environmentally aware and therefore have already lowered their discretionary electricity use. Several of these studies are designed from a psychology of behaviour perspective and disregard issues of supply and infrastructure. For example, an ethnographic analysis conducted in Oslo (Wilhite & Ling, 1992) shows that the type of electricity generation on the grid can affect consumers' attitudes towards conservation. Because Norway meets the majority of its electricity needs from hydroelectricity, homeowners rarely linked electricity consumption to environmental impacts.

What emerges is a convention in behavioural science to place individual knowledge and feedback at the centre of the equation. Individual behaviours would be caused by attitudes, and people can make choices of using or not using technologies when correctly informed. This approach has also been defined as the ABC model. A stands for individual attitudes that drive B, behaviours, because attitudes inform C, behavioural choices. According to the ABC, individuals' attitudes can be changed. For smart meters to work, the concept of choice is vital as it motivates any type of intervention form the utility, e.g. peak signals, changes in price and suggestions. According to this view, energy demand is a consequence of individual action. With better information or more appropriate incentives, individuals could reduce energy demand. The ABC model of society reduced to individual attitudes, choices and behaviours is the dominant model that informs smart-metering policy. This approach relies on psychological theory of feedback, which dates back to the 1930s, with Skinner's (1938) model of operant conditioning: behaviours which produce a positive effect are more likely to be repeated than those which produce a negative effect. According to this view, positive results could be seen as a positive reinforcement and negative results as a punishment, thus respectively encouraging or discouraging subsequent behaviour. In the 1960s, feedback was connected with goals and information about progress towards goals (Bandura, 1969). In the 1990s Feedback Intervention Theory appears and asserts that behaviour is regulated by comparisons to pre-existing goals, as behaviour is generally goal directed, and that people use feedback to evaluate their behaviour in relation to their goals. Both negative and positive feedbacks can bring about behavioural change (Vallacher & Wegner, 1987). These theories suggest that feedback is effective provided that the individual focuses on the feedback goal. This means that feedback will have to direct the attention of the individual to a specific goal compared with the *status quo*. Availability and visibility feedback are two necessary conditions for the success of feedback.

In practice

Over a hundred empirical studies of energy feedback have been conducted over the past 40 years and over 200 articles have been published about energy feedback during that time (Karlin et al., 2014). What transpires from the behavioural approach on smart meters is that the literature is mixed on the conservation effect associated with smart meters, but the most widely cited range is between 3% and 15%. The meta-analysis by Fischer (2008) found usual savings are between 5% and 12%. The Faruqui et al. (2009) meta-analysis found a range of 3%–13% and concluded that consumers who actively use a smart meter can reduce their consumption of electricity by an average of 7%. Darby (2006) reaches a similar conclusion for smart meters and claims that savings are typically of the order of 10% for relatively simple displays. An ACEEE review finds an average conservation effect of 8.6% (Ehrhardt-Martinez et al., 2010). So for smart meters' specific feedback, the literature suggests a figure of 6%–10% with some acknowledgement that larger-scale, newer studies come in towards the lower end of this range. The same ACEEE review concludes that future research should give more attention to understanding the significant variation in energy savings that has resulted from the application of real-time feedback technologies.

However, the most comprehensive meta-study of trials in this area (the AECOM trial results were released in 2011) finds only a 3% conservation effect from smart meters and warrants further examination; the AECOM report finds no statistically significant in-home display effects in three out of four studies without the presence of a smart meter.

The review in Chapter 3 of European countries shows that cost–benefit analyses have adopted very broad ranging energy savings assumptions. For example, the cost–benefit analysis for the Dutch Ministry of Economic Affairs assumes 6.4% electricity savings with direct feedback through an in-home displays (3.2% with indirect feedback) and 5.1% (3.7%) for gas. The cost–benefit analysis in Ireland adopts a 3% electricity savings assumption to compute illustrative estimates of the change in consumer welfare resulting from the installation of smart meters.

More recently, in 2015 the UK government published the findings of the Early Learning Project (ELP) – an extensive programme of research into how best to deliver consumer benefits through effective engagement. This report presents findings on GB consumers with received smart meters installed between 2011 and early 2013. The quantified impact of early smart-type meters on household energy consumption in the year following installation was in the range between 1.6% and 2.8% for electricity and between 0.9% and 2.1% for gas (DECC, 2015).

The rise and fall of the net conservation effect

Three key findings emerge from this review of smart-metering trials.

First, analysing the conservation effect from real-time feedback on a dedicated smart meter from the 1970s to present shows a steady decrease in the conservation effect achieved. Second, analysing the effect of sample size on conservation shows that as sample sizes increase, the conservation effect achieved from smart meters decreases. Third, previous studies on the effects of smart meters can be categorised by their sample selection methods and level of involvement by the study administrators. Through this categorisation, it is demonstrated that the studies with more representative sampling and recruitment methods, and lower levels of administrator involvement, are likely to exhibit lower conservation effects.

This section reviews existing meta-analyses on a large number of individual studies, and conference proceedings where results from smart meters trials have been announced by electric utilities, electricity retailers and equipment vendors (for example, Smart Grid Australia, 2011; Utilimetrics, 2011). The review also comprises press releases from smart-metering vendors and associated partner companies such as Tendril, Control4, Itron and the Zigbee Alliance (2011) regarding announcements of forthcoming smart-metering trials in Canada, the United States and Australia.

Using information from these sources, a database of 33 completed trials from 1979 is assembled with a view to understand how real-time feedback affects residential electricity consumption through smart meters.

Size matters

Can the number of participants in a trial have an impact on net conservation effects of smart meters?

On the one hand, according to some, there does not exist any correlation between the size of a trial and its outcome with regard to reported savings and awareness. A study on 12 large-scale smart meter trials with 200–2000 participants demonstrates the same range of conservation as the entire group of 38 studies comprising the same review (Darby, 2000). Another meta-analysis concedes that the scarcity of well-designed large-N studies may influence the expected levels of savings (Fischer, 2008).

On the other hand, other analysts believe size matters. The review conducted by ACEEE finds a mean of 11.6% for smaller studies, with less than 100 households, and 6.6% for larger studies, with more than 100 participants. The review concludes that large-scale smart meter trials are also more likely to experience less remarkable conservation effects. These findings seem to be in contrast with Darby's earlier findings, where she states that "When the more recent studies are compared with those carried out over the whole 25-year period, the ratio of 'successful' ones (5%+ or 10%+ savings) to the whole is almost exactly the same" (Darby, 2000, p. 8).

What can be observed without any doubts is that there has been a signifi-cant increase in sample sizes of real-time feedback trials associated with the year in which the trial was conducted. The growth in sample sizes relates to the increased budgets for smart-metering trials over the last decade. Regardless of the case of such growth in sample sizes, the results from smart-metering trials have increased in validity and clearly shown that from customer responses to these devices policy-makers cannot expect conserva-tion effects in any region higher than 3%–5%.

The trend in conservation effects achieved from real-time feedback by the year in which the trial was completed is approximatively linear over time. A remarkable decrease in conservation effect can be observed from the early studies, which typically achieved a 10%–15% reduction, compared to more recent studies that are clustered around a 5% reduction. The average across all the studies yields a conservation effect of 6.4%.

In work by McKerracher and Torriti (2013) smart meter trials are sepa-rated in terms of when they took place (from 1979 to 2004 and post-2005). The distinction of 2005 to present is to some extent arbitrary, and the data from the studies used in this analysis are not exclusively more advanced (or smart grid-enabled). Many of the newest trials are still based on clip-on current sensors (The Energy Detective) or optical sensors (PowerCost Monitor), rather than an in-home display directly paired to a smart meter over a Home Area Network. However, 2005 represents the year of substan-tial completion of the first large-scale smart grid project, carried out by Enel SpA in Italy with approximately 27 million customers. It is also the year of publication of one of the first academic papers to use the term "Smart Grid" (Massoud & Wollenburg, 2005).

Separating smart-metering trials in terms of when they took place (from 1979 to 2004 and post-2005), the eight trials from 1979 to 2004 are associ-ated with a mean conservation effect of 11.1%, whereas the 19 trials from 2005 to present have a mean conservation effect of 4.5%. By applying a weighted mean, the samples size of each study can be taken into account. The weighted mean conservation effect from a smart meter results to be 3.5% across a total of over 13,000 trial participants.

The differences between trials for the period from 2005 to present and those for the period from 1979 to 2004 can be explained in three ways. First, The AECOM trials found no statistically significant effect of Real Time Displays, efficiency advice, historic feedback, self-reading of meters or financial incentives to save energy in the absence of smart meters. They also found that the Real Time Displays that were paired to smart meters were viewed everyday by almost 40% of homeowners and that the smart meter had a lower failure rate. Second, the increasing interest in the smart grid has been a major driving factor behind the growing budgets for real-time feedback trials. For instance, the American Recovery and Reinvestment Act of 2009 provided $3.5 billion in funding for Smart Grid Investment Grants, including a substantial number of Home Area Network trials.

This funding program alone has catalysed more than 100 smart grid trials in the United States (U.S. Department of Energy, 2011). Three of the upcoming trials and one ongoing trial included in the dataset presented in this chapter received funding from the Department of Energy through this program. Smart-metering trial data from EDF, E.ON, Scottish Power and SSE, which have comparatively large sample sizes, can similarly be attributed to the UK plan to roll out smart meters to 26 million homes by 2020, which led to £9.75 million in government funding for the trials (Hamidi et al., 2009). Third, the increase in sample sizes of real-time feedback trials can essentially be associated with the higher investments in smart-metering trials. For instance, Greentech Media (2011) reports that the Home Area Network trial consisted of 20,000 households by Nevada Energy.

The conservation effect demonstrated against the trial size highlights a cluster of large-scale studies around the 1%–5% conservation mark. The data demonstrate a consistent trend: larger sample sizes are correlated with lower conservation effects. These findings by McKerracher and Torriti (2013) contradict those of Darby who claimed no correlation between study size and conservation.

Trial design matters

The causes for this decreasing trend in conservation effects might be related to what is known as "Hawthorne Effect". This is a typical drawback of experimental methods in which participants are conscious of being part of a trial. It can be explained in terms of effects associated with study participants being excessively aware of being monitored. This means that any measured change is a consequence of the study design and of the behavioural intervention being tested. In principle, the Hawthorne Effect should be more visible in smaller studies because these generally involve a higher level of contact and interaction between study administrators and participants. Large smart meter trials that involve minimal intervention such as the AECOM results and the Western Power trial are less likely to generate results which are affected by the Hawthorne Effect. Scottish Power used balanced trial groups which were designed according to their level of electricity consumption. These customers were not aware that they were participating in a smart-metering trial, though both suitable wiring at the meter and customer consent were pre-requisites for the installation of the clip-on displays.

Sampling matters

The methodology for sampling and recruiting respondents is a significant factor explaining variation in the findings of large studies regardless of the fields in which they are conducted. The same should apply to the methodology for sampling and recruiting households in smart meter trials and how these might impact their savings effects. The majority of trials (Mountain, 2008; Sulyma et al., 2008) do not specify their sampling and recruiting methods, whereas the review studies generally do not include these as explanatory variables. On the contrary, the ACEEE meta-analysis provides a simple taxonomy of programs: "opt-in" and "opt-out". It transpires that the former largely realise participation rates in the range of 75%–85%, while the latter achieves less than 10%. However, these (very differing) participation rates are not linked to saving effects.

Since the way participants are selected and recruited for trials inevitably has an impact on the wider applicability of their findings. For example, a study in California looking at the type of conservation behaviour households engaged in as a result of feedback found that 50% of the pilot participants had $100,000 or more in annual income, compared with 12% of the general population. Very small studies are particularly affected by less representative sampling methods. In a 10-person study, findings show that participants who volunteered had generally already engaged in more one-time efficiency improvements (Allen & Janda, 2006).

Findings from McKerracher and Torriti (2012) show that the use of unconventional sampling techniques or sampling bias boost conservation effects. This occurs especially where participants may already have prior motivation and knowledge about energy saving possibilities which can be activated by feedback. There are exceptions, such as a Japanese trial (Ueno et al., 2006) which deploys representative sampling and recruitment methods and yet shows high conservation effects. This might also be explained by the involvement of a very elaborate and expensive display, which likely improved the conservation effect.

The studies outlined below used sampling techniques which can be regarded as some of the most representative examples of recruitment of customers in smart-metering trials to date.

In their UK trials, Electricite de France made use of stratification by levels of electricity consumption (low, medium and high). Random number generation was used to allocate trial groups to households. The recruitment in the trials by Electricite de France were carried out based on the principle of customer awareness. There was an opt-in option, and all households participating in the trial were informed before the study began. Electricite de France recruited end users by informing them that they had been temporarily allocated to a study. Subsequently, this pool of customers was asked to take part in the trial by suggesting that this would bring about more precise billing, environmental benefits and reductions to bill costs.

E.ON adopted random recruitment for selecting its sample population also by introducing quotas for fuel poor and non-fuel poor homeowners. With regard to clip-on devices, end users were not invited to be part of a trial and were not informed that they would be taking part in a trial, despite the fact that an installer visit was required. With regard to smart meters, the recruitment had an opt-in basis. However, participants were not informed that they would be participating in a trial and E.ON achieved participation rates closer to an opt-out study by using the requirement for "meter re-certification" as an argument for households to participate.

As a result, the most important message from these trials with representative sampling is that on average only 3% conservation effects were achieved attributable directly to the smart meter. The most representative sampling and recruitment techniques were used in the Scottish Power trial. The most significant trial achieved the lowest reduction of the four. Moreover, in the E.ON and SSE studies, clip-on current sensor versions were associated with the most representative sampling and recruitment techniques, and these on average achieved only a 1% reduction.

Other issues which matter when considering net conservation effects of smart meters

There are at least three additional issues which emerge from this review and should be taken into account when estimating the expected conservation effect of smart meters.

First, existing meta-analyses point to no clear indication about whether long-term projects provide higher initial savings than short-term ones (Fischer, 2008). However, concepts of fatigue are highlighted in the theoretical ABC literature as potentially diminishing conservation effects of smart meters in the medium and long run. Empirical evidence from the meta-analysis presented in this chapter shows that the length of trial has a mixed effect on net conservation from smart meters. Similarly, the ACEEE review found savings from feedback in shorter studies of six months or less to be 10.1%, while studies over six months achieved only 4.7% reduction. They attribute this to conducting shorter duration studies in the summer months. Increased duration leads to increased persistence of conservation effects (Darby, 2008).

Second, there is an expanding literature on most effective ways to communicate data to consumers (Ai He et al., 2010; Wilhite et al., 1999). There might be some merit in confronting the results on the basis of the fact that growing competition in the smart-metering industry and improvement in user interface are likely to change the way communication is portrayed on the digital meter and transmitted to users.

Third, the implementation of further hardware in the smart grid may result in decreases in drop rate for smart-metering trials because clip-on current smart meters require an installation process, while more advanced smart meters do not require one. As part of one of the AECOM trials, findings show that in an early survey of households which were sent the clip-on meter, 46% had received and fitted it and only 31% were still using it. This problem of homeowners not using the device or not activating it can affect smart meter-based trials. For example, in a 2,200-unit smart meter trial in Western Australia, advanced smart meters were sent out randomly to homes within the selected trial sample without prior knowledge or consent from the home-owners. The package that was sent included marketing material explaining how to commission the device and deploy it. Despite the intended uncomplicatedness of this process, only half of participants activated their smart meters. Measuring the anticipated drop rate and how it will be affected by using different typologies of smart meters is not a straightforward thing to do. In general, the existing trials and some common sense suggest that smart meters will have less impact if customers are relied upon to fit them.

Disaggregated feedback provides information on the breakdown of how electricity is consumed within the home. This can take place either using individual end-use plug monitors or having the in-home display detecting different load signatures of various home appliances. Similarly, display devices can also provide the homeowner with some control capabilities, such as being able to switch on or off certain electrical loads. Such devices are typically more expensive than basic display devices and rely on built-in Home Area Network communication capability from appliances. For this reason, disaggregated feedback and control capability are not covered in depth in this chapter.

Prepayment for electricity has also been demonstrated to reduce residential consumption, with keypad displays also providing regular updates on usage. One trial in Ireland found the conservation effect from this approach to be around 11% (Faruqui et al., 2009). However, prepayment involves different aspects of the consumption decision than real-time feedback alone and is therefore not treated in depth here.

Conclusions on net conservation effects

Based on the information acquired in this review, 4% is the most reliable available estimate of the expected conservation effect from smart meter feedback.

This chapter set out to shed light on some confusion in the literature with regard to the net conservation effects from smart meters. Theoretical imperatives and empirical discrepancies dominate any discussion on the effectiveness of smart meters in reducing energy demand. Theories which put the individual at the centre of decisions on energy consumption embrace the idea that significant conservation effects can originate from direct feedback. This is in tune with approaches in psychology which value direct feedback as a determinant of goal-oriented behaviour. In empirical studies the most widely cited range for conservation effects from direct feedback is between 3% and 15%. In her meta-analysis, Fischer (2008) finds that savings are typically between 5% and 12%. Investigating feedback from smart meters, the Faruqui et al. (2009) meta-analysis find that on average consumers who actively use a smart meter can reduce their consumption of electricity by 7%, with a range of 3%–13%. In her analysis Darby (2006) concludes that savings are typically of the order of 10% for smart meters with relatively simple displays.

The ACEEE and Stromback et al. (2011) studies find very similar average conservation effects of 8.6%–8.7%, respectively. Overall, it can be concluded that for smart meters and direct feedback, the review of empirical studies suggests conservation effects of around 6%–10%. In the literature it is often mentioned that larger-scale, newer studies are closer to the 6% rather than the 10%. An example of this comes from the AECOM trial in 2011, which concludes that the conservation effect from smart meters was in the order of 3%.

References

AECOM. (2011). *Energy demand research project: Final analysis.* Hertfordshire: AECOM House- Prepared for Ofgem.

Ai He, H., Greenberg, S., & Huang, E. (2010). One size does not fit all: Applying the transtheoretical model to energy feedback technology design. In *CHI '10-Proceedings, 28th International Conference on Human Factors in Computing Systems*, Portland, OR.

Allen, D., & Janda, K. (2006). The effects of household characteristics and energy use consciousness on the effectiveness of real-time energy use feedback: A pilot study. *ACEE Summer Study on Energy Efficiency in Buildings.* Washington, DC: American Council for an Energy Efficient Economy.

Bandura, A. (1969). *Principles of behaviour modification.* New York: Hold, Rinehart & Winston.

Brandon, G., & Lewis, A. (1999). Reducing household energy consumption: A qualitative and quantitative field study. *Journal of Environmental Psychology, 19*, 75–85.

Cunningham, W., & Joseph, B. (1978). Energy conservation, price increases and payback periods. In H. K. Hunt (ed.), *Advances in Consumer Research* (pp. 201–205) 5. Ann Arbor, MI: University of Michigan.

Darby, S. (2000). Making it obvious: Designing feedback into energy consumption. *Proceedings, 2nd International Conference on Energy Efficiency in Household Appliances and Lighting.* Italian Association of Energy Economists/ EC-SAVE programme.

Darby, S. (2006). The effectiveness of feedback on residential energy consumption. A review for DEFRA of the literature on metering, billing and direct displays. Environmental Change Institute, University of Oxford.

Darby, S. (2008). Energy feedback in buildings–Improving the infrastructure for demand reduction. *Building Research and Information, 36*, 499–508.

Darby, S. (2010). Smart Metering: What potential for householder engagement? *Building Research and Information, 38*, 442–457.

DECC. (2015). Smart metering early learning project: Synthesis report. Oxford: The University of Ulster, and the Tavistock Institute

Ehrhardt-Martinez, K., Donnelly, K., & Laitner, S. (2010). *Advanced metering initiatives and residential feedback programs: A meta-review for household electricity saving opportunities.* Washington, DC: American Council for an Energy-Efficient Economy (ACEEE).

Faruqui, A., Sergici, S., & Sharif, A. (2009). The impact of informational feedback on energy consumption- A survey of the experimental evidence. Washington, DC:

The Brattle Group. Retrieved from www.brattle.com/_documents/uploadlibrary/ upload772.pdf

Fischer, C. (2008). Feedback on household energy consumption: A tool for saving energy. *Energy Efficiency, 1*, 79–104.

Greentech Media. (2011). NV energy picks Control4 for residential demand response. Retrieved from http://www.greentechmedia.com/articles/read/ nv-energy-picks-control4-for-residential-demand-reponse/

Hamidi, V., Li, F., and Robinson, F. (2009). Demand response in the uk's domestic sector. Electric Power Systems Research, 79 (12):1722–1726.

Hutton, B., Mauser, G., Filiatrault, P., & Ahtola, O. (1986). Effects of cost related feedback on consumer knowledge and consumption behaviour: A field experimental approach. *Journal of Consumer Research, 13*, 327–336.

Karlin, B., Ford, R., & Squiers, C. (2014). Energy feedback technology: A review and taxonomy of products and platforms. *Energy Efficiency, 7*(3), 377–399.

Massoud, A. S., & Wollenburg, B. (2005). Toward a smart grid: Power delivery for the 21st century. *IEEE Power and Energy Magazine, 3*, 34–41.

McKerracher, C., & Torriti, J. (2013) Energy consumption feedback in perspective: Integrating Australian data to meta-analyses on in home displays. *Energy Efficiency, 6*(2). pp. 387–405.

Mountain, D. (2008). *Real-time feedback and residential electricity consumption: British Columbia and Newfoundland and Labrador pilot.* Ontario, ON: Mountain Economic Consulting and Associates.

Nielsen, L. (1993). How to get the birds in the bush into your hand. Results from a Danish research project on electricity savings. *Energy Policy, 21*, 1133–1144.

Parker, D., Hoak, D., Meier, A., & Brown, R. (2006). How much energy are we using? Potential of residential energy demand feedback devices. *Proceedings of the ACEEE 2006 Summer Study on Energy Efficiency in Buildings*, vol. 1, pp. 211–222.

Skinner, B. F. (1938). *The behavior of organisms: An experimental analysis.* New York: Appleton-Century.

Smart Grid Australia. (2011). Smart grid Australia news. Retrieved from http:// www.smartgridaustralia.com.au.

Stromback, J., Dromacque, C., & Jassin, M. (2001). The potential of smart meter enabled programs to increase energy and system efficiency: A mass pilot comparison. Report prepared by VaasaETT Global Energy Think Tank.

Sulyma, I., Tiedemann, K., Pedersen, M., Rebman, M., Yu, M. (2008). Experimental evidence: A residential time of use pilot. *Proceedings of the ACEEE 2008 Summer Study on Energy Efficiency in Buildings*, 2 p. 292–304.

Torriti, J., Hassan, M. G., & Leach, M. (2010). Demand response experience in Europe: Policies, programmes and implementation. *Energy, 35*, 1575–1583.

Ueno, T., Sano, F., Saeki, O., & Tsuji, K. (2006). Effectiveness of an energy consumption information system on energy savings in residential houses based on monitored data. *Applied Energy, 83*, 166–183.

United States Department of Energy. (2011). Department of energy-smart grid investment grant awards. Retrieved from http://www.oe.energy.gov/recovery/1249. htm.

Utilimetrics. (2011). Utilimetrics autovation- the smart utility conference and exposition. Retrieved from https://www.businesswire.com/news/home/20090519005514/ en/Utilimetrics-Autovation%C2%AE-2009-Conference-Held-Denver.

Vallacher, R., & Wegner, D. (1987) What do people think they are doing? The presentation of self through action identification. *Pyschological Review, 94*(1), 3–15.

Wilhite, H., Hoivik, A., & Olsen, J. (1999). Advances in the use of consumption feedback information in energy billing: The experiences of a Norwegian energy utility. *Proceedings, European Council for an Energy-Efficient Economy/Panel III*, 02.

Wilhite, H., & Ling, R. (1992). The person behind the meter: An ethnographic analysis of residential energy consumption in Oslo, Norway, ACEEE 1999 Summer Study on Energy Efficiency in Buildings, Washington, DC: American Council for an Energy Efficient Economy.

Zigbee Alliance. (2011). Zibgee smart energy overview. Retrieved from http://www.zigbee.org/Standards/ZigBeeSmartEnergy/Overview.aspx.

7 Beyond economics and system needs

Theoretical approaches to understand smart meters energy demand

Understanding the timing of energy demand

Understanding energy demand involves analysis over individual behaviours related to energy, the structure of everyday life, how practices are sequenced throughout the day and how energy demand varies in time and space. This is because the timing of people's social practices throughout the day determines when demand is either high or low.

For some time, the main metric for analysing the structure of energy demand has relied on average load profiles on a daily or monthly basis. Electricity billing is typically carried out at a low time resolution, either on a monthly or bi-monthly basis. Key figures on annual consumption in a country or providing comparisons across countries tend to rely on averages. More recently, increasing concerns over demand and supply balancing in association with high levels of integration of renewables have pushed analysts and researchers to move beyond average and think more systematically about the structure of energy demand with higher time resolution (e.g. half hours or even minutes). The timing of lighting and heating, for example, depends on the timing of what people do when they are home and influences not only the individual electricity consumption profile in a household, but also demand in a neighbourhood or district (i.e. distribution network); and increases in demand in the transmission grid. When demand is structurally high and causes peaks, typically in the late afternoon of a winter day, the costs and negative environmental impacts of meeting this extraordinarily high demand are higher than normal.

How can energy demand be analysed in terms of its structure? Different approaches have been used, from stochastic reproductions of appliance use to deterministic representations of load profiles. A variety of conceptual positions have been used to explain the structure and shape of energy demand. These include behavioural theories, which deal with attitudes, preferences and choices around energy consumption, and social practice theories, which address issues of timing, space, rhythms and societal change in energy demand. The main contribution of this chapter consists of looking at both approaches in terms of their theoretical propositions, but also with regard to their empirical research applications.

Carrying out analysis on the timing of energy demand involves understanding when energy demand occurs (i.e. at what time of the day). This is inextricably related to questions of why energy demand takes place at specific times of the day. Smart meters can address questions regarding how much demand occurs at different temporalities, but they cannot address questions around the reasons why energy is used in the first place, i.e. how demand is structured.

Any analysis of the timing of energy demand needs to take into account people's activities and practices as the main variable. A simple example clarifies the thinking behind this position and is based on the substantial difference between residential electricity load curves for weekdays and weekends. During the same season, the weather can be equal at the weekend compared with the weekday. Building type, appliances, fuel substitution, price of energy and appliance control, and the moment of the day in which sunlight is present or absent may be the same between weekday and weekend. The only substantial change between weekday and weekend is in terms of people's activities.

This chapter provides background on people's activities and the timing of energy demand; postulates key concepts with particular emphasis on behavioural theories and social practice theories; presents applications, for instance, by reviewing behavioural theories applied to smart meters and social practice theories applied to time use studies); provides observations on applying behavioural theories and social practice theory in the context of smart metering; and discusses the overall role of the social sciences with regard to smart meters.

Key concepts to understand smart meter energy demand data: behavioural theories

What is the role of behavioural theories within the context of smart meter energy consumption data? The starting point of this approach is that human behaviour plays an important role in the way energy is consumed. Behavioural determinants that influence energy consumption are shaped by internal (such as individual attitudes, beliefs, values, personal norms) and external factors (such as regulations, institutional constraints and consumption practices) (Stern, 2000). In order to understand and to change energy consumption behaviour, both internal and external influences need to be taken into consideration.

In the past ten years, the number of behavioural models has continued to increase, mainly due to two reasons: first, due to the complexity of the cognitive processes underlying behaviour, and second, there is no single unified approach to understanding behaviour.

The strength of behavioural models is in their ability to represent factors affecting individual energy consumption behaviour (e.g. knowledge, technology, etc.) that are important indicators for policy-makers in designing behaviour change policies. When applied to smart meters and energy consumption, the behavioural models have been criticised for their unrealistic assumptions around people decision process, assuming that higher knowledge leads to more positive attitudes and more self-efficacy which result in the desired action (Hargreaves et al., 2013). While electricity bills and smart meters make electricity consumption visible, several researchers suggest that technology alone is not enough to change energy consumption behaviour. Without smart meters, electricity is not visible to us, but it is so embedded in our everyday routines that it is difficult for people to connect to a specific behaviour (Hargreaves et al., 2010). Strengers' (2013) image of the 'Homo Economicus' or 'Resource Man' captures this ideal type of smart energy consumer, seen as being a technologically minded, information-oriented individual who actively monitors his environment using accurate and up-to-date energy information about the costs, resource units (kWh) and impacts (CO_2) of his consumption ensuring that he is not at risk of 'illness' or 'diseases' caused by his energy behaviour. His smartness is defined by his engagement with the data and technology. On the other hand, the 'Resource Man' conceptualisation ignores what individuals actually do in their homes or the way electricity is shared among household members. Furthermore, it offers a limited understanding of any change in electricity consumption.

In summary, behavioural models have been applied to understand and to predict energy consumption intentions, food choice or travel mode choice. Behavioural models are useful in identifying people's motivations in terms of adopting and using smart meters. Motivations include environmental concerns, better comfort and control over electricity bills. Two models are discussed below as they are frequently used in smart meter research: the Attitude-Behaviour-Context model and the Theory of Interpersonal Behaviour.

The attitude–behaviour context (ABC)

Behaviours and actions around smart meters are shaped not only by interpersonal determinants but also by contextual factors such as government regulations or price incentives. Stern (2000) in developing the ABC model focused on understanding the relation and structural dynamics between attitudes and contextual factors. Energy behaviour in this model is the product of internal attitudes towards efficiency and conservation and external contextual factors, such as feedback from smart meters.

Beliefs, norms, values and a tendency to act in certain ways shape attitudinal variables (Jackson, 2005). A literature review on predictors of energy-saving intention (Guerin et al., 2000) suggests that attitudes are strong predictors of energy-saving intention, but occupant characteristics (such as gender, age, residency) are better predictors of specific energy consumption actions. Contextual factors (e.g. information from smart meters) can play a major role in supporting personal attitudes in performing energy consumption behaviour. For instance, consumers who have greater trust in information provided by governments, environmental agencies and friends/family are more likely to rely on eco-labels in their product purchases.

The strength of the contextual factors determines the extent to which personal factors affect any particular behaviour, thereby creating a hierarchy of influences. Attitude–behaviour relationships are strongest when the external context provides weak pressures for or against any particular behaviour (Stern, 2000).

Theory of interpersonal behaviour

Energy consumption behaviour is influenced by personal (e.g. attitudes, values, habits and personal norms) as well as by wider societal factors (e.g. institutional constraints and social practices). Behavioural theories cover not only the cognitive variables affecting the energy-related decision process (e.g. reading a smart meter and acting), but also the wider social and institutional context within which energy is produced, distributed and consumed.

According to the Theory of Interpersonal Behaviour, individuals' energy consumption behaviour is either the outcome of rational decision-making processes or a habitual process based on the frequency of past behaviour. In the first case, energy behaviour is determined by social and emotional factors as well as by the consequences associated with performing that specific behaviour (e.g. smart meters prompting individuals to turn off the lights in order to save money). In the second case, energy consumption is conceptualised as the outcome of a habitual action that is relevant for domestic energy consumption.

This approach provides a comprehensive view of the structural complexities involved in behavioural research and the determinants that influence behaviour, suggesting behaviour change strategies. However, according to this approach, prediction is difficult when the individual is not highly committed to a certain intention.

Key concepts to understand smart meter energy demand data: social practice theories

The conceptualisation of the energy consumer as a rational actor who uses smart-metering technologies to manage demand generates limitations in understanding how demand is constituted and changing. This calls for an investigation into everyday energy consumption routines to find out how and why householders consume electricity in their everyday lives. Rather than focusing on individual behaviours, energy consumption can be defined as the outcome of people's everyday doings and sayings (Warde, 2005). This suggests that individuals do not consume electricity for the sake of it, but in order to carry out everyday tasks in everyday life. Thus, it is within the everyday tasks, routines and rhythms that energy consumption takes place, from getting up every morning, having breakfast, going to work or school, having lunch, going home, having dinner, reading a book, surfing on the internet or watching television and probably doing similar things again and again.

The focus shifts here from behaviour to people's practices. Humans are conceptualised as carriers of practices operating within a socio-technical landscape (Warde, 2005). Detailed understanding of the way humans become carriers of practices and what their relationship with practices is could help in understanding the differences between practice theory and behavioural approaches.

Whilst in behavioural approaches the individuals and their attitudes, behaviours and choices form the basis of action, in practice theory, the people are seen as knowledgeable and skilled 'carriers' of practice who at once follow the rules, norms and regulations that hold practices together, but also, through their active and always localised performance of practices, 'improvise and creatively reproduce and transform them' (Wilhite, 2012, p. 8). Practitioners are seen as powerful agents who have the skills to creatively engage, transform, or ignore performances of practice.

If in behavioural approaches the understanding of individuals is related to a functional context, such as smart meters, and professions, in practice theory the individual is capable of being a carrier of multiple practices, or bodies of knowing, thinking or acting in multiple contexts. Practices recruit actors (and not *vice versa*). This means that smart meters alone cannot impose the new practice of continuously reading electricity consumption levels. However, through the continuous performance of practices, actors become skilled over time. This means that the way in which people interact with smart meters may change over time depending on the extent to which this becomes embedded in everyday lives. According to Warde (2005), the process of enrolment starts with domestic practices during infancy through formal associations, and social or recreational activities. In essence, if my son interacted every day with the smart meter (see Chapter 1) and acquired the knowledge and skills, along with other children of his time, the practice could evolve. However, the way in which practices are shared and transmitted from one carrier to another is subject to interpretation, but one way in which new recruits are initiated into practices is through social places. History and location are some examples of important social places (or networks). In their example on showering, Shove et al. (2012) describe the ways showering on a daily basis spread and become embedded through the design of the bathroom and expectations of friends and family. This suggests that it is possible to study smart meter utilisation through networks, infrastructures, tools and rules, but when it comes to the practical know-how, it is more challenging to understand the recruitment process.

In the case of smart meters, what is the meaning of carriers in the careers of the practices themselves? The careers of practices are responsible for the trajectory of the practice as whole. In principle, a high level of expertise may allow practices to develop and evolve, resulting in a situation in which experienced practitioners define career paths that others follow. There are cases when the aspirations within the practices are contested and practitioners remove the original system, by revolutionising a practice (Warde, 2005). For instance, renewable energy systems (such as wind turbines or solar panels) continued to revolutionise the practices-that-consume-energy (Strengers, 2013). This suggests that practices-that-consume-energy are undergoing a process of 'remediation', whereby the practices are transformed as a result of changes in the technology (Shove et al., 2012).

Interaction with smart meters: practices as entity?

Two common definitions of practices consist of practice-as-entity and practice-as-performance. A practice-as-performance takes place at a particular time and space when understandings, technologies, practitioners and activities come together in a specific way (Schatzki, 1996). The first example is that at 7.00 pm on 15 October 2019, I submitted an electricity meter reading using the in-home display. Active and constant reproduction, undertaken by multiple practitioners in diverse spaces and times, can be conceptually brought together in considering a practice-as-entity (Shove et al., 2012). In this case, the entity in question could be the practice of submitting meter readings, or more specifically the practice of monitoring home electricity usage or even of using an in-home display.

A focus on performances may reveal the variation and potential transformation of the electricity meter reading activity. According to the example, even if I submit regular meter readings (once a month), this activity precludes an exact duplication. Despite all efforts, I may never submit my electricity meter reading at exactly the same time (such as 7.00 p.m.) and in exactly the same order. There may be disruptions, such as being on holiday or receiving a text message in the middle of the reading submission activity, that transform the performance of the meter reading process. The practice of submitting regular meter readings consists of multiple non-dependent and recognisable activities (such as monitoring electricity consumption and using an in-home display). Together these activities form the entity of submitting regular meter readings.

According to the widely cited definition, a practice:

> ...is a routinized type of behaviour which consists of several elements, interconnected to one another: forms of bodily activities, forms of mental activities, forms of mental activities, 'things' and their use, background knowledge in the form of understanding, know-how, states of emotion and motivational knowledge. A practice - a way of cooking, of consuming, or working, of investigating, of taking care of oneself or of others, etc. -forms so to speak a 'block' whose existence necessarily

depends on the existence and specific interconnectedness of these elements and which cannot be reduced to any one of those single elements.
(Reckwitz, 2002, p. 250)

Practices can be defined as 'temporally unfolding and spatially dispersed nexus of doings and sayings' (Schatzki, 1996, p. 89). Practices are performed in time and space, and this is another important feature of practices. According to this view, what makes up a practice is best considered when investigating the relation of practices to each other. For example, when individuals feel cold, they draw on their practical understanding in deciding what to do. Smart meter readings may compete with other practices such as putting on a jumper, shutting blinds, turning on the heater or taking a hot shower. The practical knowledge in which smart meters may enter is not something given at birth. It is a product of social history, such as education, or social experience. For example, social expectations of appearance, smell or hygiene inform how and when comfort and cleanliness practices should be undertaken, and this might be in contrast with information on changes in tariffs as communicated by the smart meter. Rules link the doings and sayings of a practice and can influence the future course of activity. Rules or recommendations are introduced by those with the authority to reconfigure or reinforce a practice or specific components of it. For instance, recommendations about how to save energy are given by utility companies and governments in pricing rules for electricity.

Rules can also be incorporated into things through building standards, or shower timers. According to Schatzki, a practice always displays a range of tasks in order to realise projects with specific ends. Thus, while practices are learnt, the way learning takes place is through doing and mental or affective performance. For example, actors can turn down domestic appliances in managing electricity with a feeling of obligation, but without any interest in reducing consumption. The interest can, however, lie in the project of showing examples to others on saving electricity, with the end of reducing electricity bill.

In summary, disaggregating practices into elements is useful in guiding empirical inquiries, in particular in zooming in (Nicolini, 2012) on the dynamics between elements of practice, how they collectively or separately interact with other practices that shape the performance of practice over time and space or how technologies are used. The main limitation of applying this approach to the field of smart meters is that it may lead to oversimplification or predefinition of a practice.

Interaction with smart meters: practices as performance?

Practice-as-entity focuses on the structured organisation of the practice. On the contrary, the analysis of the practice-as-performance relates to the moments of integration that occur when practices are in action. Thus, when a practice such as bathing, laundering or hoovering is performed, its elements connect in different ways and it is thought that the successive reproduction of the connection between the elements of practices is the fundamental structure of the practices that drives everyday energy use (Nicolini, 2012).

A practical example, in this sense, consists of the indoor household cleanliness routines consisting not only of a 'nexus' (Schatzki, 1996) or 'specific interconnectedness' (Reckwitz, 2002) of inconspicuous practices of consumption, such as showering, laundering or cooking, but also of ways of performing (or doing) these everyday cleanliness activities. In this sense, the introduction of a new 'institutional rule' or 'technology' is influential for the practice itself only if it is captured and used in the reproduction of the practice over time. To give a practical illustration and an example of how the structure of energy demand changes over time, showering became a common practice long after the mass installation of the electric showers in the 1970s (Shove, 2003). At the beginning, showering was thought to be dangerous for the skin, particularly for women (Southerton et al., 2004), but with changing ideas about the body, regular showering became a normalised practice. As such showering only existed as a recognisable conjunction of elements, but through regular performance, the connections between elements that constitute the showering practice sustain and legitimate showering practices (Shove, 2003). The performance of practices differs from each other as the elements of the practice are always brought together differently. Thus, the practices are sustained, changed or destroyed as the connections between them are created, maintained or broken (Shove et al., 2012). Similarly, the connections between know-how, institutional knowledge, engagements and technology are made or broken through successive moments of performances, so the practice advances or is 'fossilised'. The latter is defined by Shove (2012) as a process whereby practices disintegrate when connections are no longer sustained. One of the reasons for the changing nature of the performance of the practice is that practices are inherently social in nature as participants learn practices from each other; therefore, new standards or norms emerge. Thus, as participants in social practice discuss the practice with each other, they identify the correct ways of undertaking it and change their routines to either conform to or deviate from this new understanding. Therefore, different configurations of practice components lead to change.

Can smart meters interact with practices as performance? When existing connections break and shift, or the elements become unstable, innovations may occur and practices and elements of practices may be reinvented (Shove et al., 2012). 'Breaks' and 'shifts' in the reproduction of practices emerge from 'everyday crises of routines, in constellations of interpretative indeterminacy and of the inadequacy of knowledge with which the agent, carrying out the practice, is confronted in the face of the "situation"' (Reckwitz, 2002, p. 255). Situations leading to household crises of routines are moving home (Foulds et al., 2014), power blackouts (Rinkinen, 2013), a new household member, introduction of a new appliance or illness in the household.

In summary, the literature used above defines practices as enduring and recursive entities reproduced through recurrent performances. The transformation of practices is a dynamic process involving shifts and breaks in their reproduction. Importantly, complexes of practices exist only because the connections between different elements are consistently reproduced in 'timespace' (Schatzki, 2010).

Can smart meters assist in tracing practices in time and space?

As mentioned in the previous section, practices do not exist in isolation, but rather form a nexus within a particular site that is held together by practices, artefacts and spaces (Nicolini, 2012). For example, doing the laundry consists of several individual practices, such as selecting the laundry, loading the washing machine, drying the clothes, ironing and storing laundered clothes and others. While these are separate practices, they are usually bundled together when performing the laundry practice. More generally, as elements connect together to create a social practice, practices connect together to create bundles that are 'loose-knit patterns based on the co-location and co-existence of practices (Shove et al., 2012, p. 81)' and complexes that are 'stickier and more integrated combinations'. When practices are bundled, they can share elements and co-evolve. For example, with the evolvement of wireless technology, the use of laptops, tablets and smart phones became less spatially constrained. As the outcome of the wireless connectivity, bundles of household practices-that-consume-electricity, such as heating, changed. Thus, if in the past the whole family might spend the evening together in the living room, only heating one room, today family members could be dispersed around the house watching or communicating, using multiple electronic devices while heating the whole house.

The extent to which smart meters could be used by social scientists as a tool for understanding how bound together practices are depends on the way in which practices boundaries can be identified. One way is to classify practices as dispersed or integrative (Schatzki, 1996; Warde, 2005). According to Schatzki (1996, p. 92) dispersed practices appear in many different contexts of social life, and their performance requires the 'know how' and 'understanding' of how to carry out an appropriate practice. Examples of dispersed practices are turning on and off thermostats and using remote controls. Integrative practices are 'the more complex practices found in, and constitutive of, particular domains of social life' (Schatzki, 1996, p. 98). Examples of integrated practices with relation to energy consumption in households are cooking, laundering and heating.

According to this basic classification, smart meter reading would probably be bound to dispersed practices. Could smart meters be seen as part of integrated practices instead? Schatzki defines several ways of thinking of an integrative practice; three propositions are important in this case. The first proposition is that the performance can be read and judged as correct or acceptable (Schatzki, 1996, pp. 100–102). The second proposition suggests that integrative and dispersed practices are socially organised and coordinated bundles of activities or in Schatzki's (1996, p. 105) words 'practices are social, above all because participating in them entails immersion in an extension tissue of coexistence with indefinitely many other people'. Integrative practices are more complex partly because of the pressure exerted by the four components of the practice functioning with their 'own logic and its own different coordinating agents and organisations (Warde, 2013, p. 25)'. Coordinating agents are responsible for the interaction and communication between practices and can be identified as nodes in a socio-material network (Vihalemm et al., 2015, p. 40). One of the benefits of this network map is visualising the variation in the density of relationships between practices. Furthermore, a network map often suggests boundaries between different practices because there is a high density of interrelationship between some components forming distinct nodes.

Smart meters and the organisation of practices in time and space

Smart meters provide objective measurements not only of volumes of demand (in kWh) but also of time (with 30-minute measurements of consumption and associated variations) and space (i.e. taking the home as a unit). Practices are situated in objective time and space and result from particular historical, cultural and social conditions (Schatzki, 2009; Shove, 2012; Southerton, 2006). Time and space are not separable matters and are central to the organisation of practices. Time and space are constituted by practices, meaning that the patterns of time and space are reproduced in everyday life through participation in different practices (Shove et al., 2012). An individual follows a path in time and space, participating in practices and projects that leave 'sediments' in their minds and bodies, which support participation in certain practices at the expense of others (Røpke, 2009). In this representation, the practitioners' time is a limited, finite resource and its allocation 'reflects the relative dominance of some practice over others' (Shove et al., 2012, p. 127). This means that the dominant practices are those which consume most of the time and their dominance and continuation depend on how practitioners spend their days. Practices thus compete for time, but from the point of view of the practitioner enacting a practice properly, it is 'a matter of weaving it into an existing rhythm' (Shove et al., 2012, p. 127). Although in theory everyone has access to the same amount of time, in fact practitioners have limited amounts of time free from their scheduled activities (such as work practices), meaning that the new practices need to be inserted within these temporal rhythms or have the ability to push aside established practices (Røpke, 2009). For the same reason, any information originating from smart meters will reflect conflicting times of those who are at home. Southerton (2006) has proposed that social practices operate along Fine's (1990) five dimensions of time: tempo (or speed), duration, timing (synchronisation), sequences (ordering of events) and periodicity (rhythm). The temporal rhythm of the day is predictable and manageable by practices, which holds a fixed position in the sequence of practices (Southerton, 2006, p. 436). In relation to energy consumption life events such as house moves (Foulds et al., 2015; Gnoth, 2013), home renovations (Energy Saving Trust, 2011) or changing family roles (Shirani et al., 2013) can create conditions where individuals or households are more receptive to an adjustment in their energy-related practices. In this way, life events may represent a window of opportunity for inducing change in practices-that-consume-electricity. The extent to which technologies such as smart meters can feature in these moments of change is debatable (Burningham et al., 2014). While the conceptualisation of practices as competitors in time and space entails that practices require recurrent performance, practices are rarely continuous. Rather, practices are active and at other times 'lie dormant' (Shove et al., 2012, p. 128). This suggests that the elements of the practice must be 'lying around' and waiting to be integrated in a practice to be able to bridge between performance moments. Seen from this perspective, time and space have a dual role: to connect performances in objective time and 'move structuring in the way in which time is experienced and organized within society' (Shove et al., 2012, p. 129). This leaves the question of how dormant practices continue to exist or how they can be revived at different moments in both time and space. Commonly accepted approaches to time and practice frame practices as 'consumers of resources' (Shove et al., 2009) that are competing with each other for practitioner's attention. According to this view, smart meters could be seen as proxies, revealing consumption not only of energy but also of time dedicated to practices. However, it is possible to conceptualise time as the product of practices that is an 'integral part of a systems of practice in which we are all engaged' (Shove et al., 2009, p. 24). In this framing time is integral part of practice; in other words, it takes certain time to accomplish properly a practice that would suggest a zooming in (Nicolini, 2012) on how everyday practices are the makers of the time (Røpke, 2009). According to this view, smart meters lose their attribute of proxies/practice-revealers.

In order to understand how smart meters reveal everyday electricity consumption, it is important not only to engage with practices-that-consume-electricity from a macro perspective but also to examine how electricity is experienced in the living of everyday life. Schatzki makes distinction between the experienced time and space (so called timespace) and also 'space-time' which stands for objective space and time. The difference between the two is that 'Human activity...institutes and bears a timespace whose temporal and spatial dimensions are connected inherently – and not contingently as with objective space-times' (Schatzki, 2010, p. 38). In other words, time and space come together through action and are derived from 'time-space actions and from the practices that actions compose'. Schatzki argues further that 'an important feature of temporality as a feature of human activity is its relation to teleology' (Schatzki, 2010, p. 28). Thus, the teleological characteristics of time suggest a strong orientation towards all kinds of ends. In this sense, the 'timespace' (Schatzki, 2010) of human activity is defined as 'acting toward ends from what motivates at arrays of places and paths anchored at entities' (Schatzki, 2010, p. 38).

Schatzki's definition of 'timespace' could help in understanding practitioners' embodied actions as mutually constituting domestic 'timespace'. For example, the practice of cleaning the bathroom is a routine process through which the rhythms of renewal and re-growth are regulated (Pink, 2012). Both the connections between the growth of limescale and accumulation of dirt and the performance of cleaning define the orientation of the practice, and it gives its timing a moral meaning. The relationship between the performance of cleaning and the phenomenology of times is also significant in that 'cleaning time' is made meaningful (Pink, 2012) through the listening of radio or music, which in itself contributes to the sensory framing of the timespace of cleaning. This way the moral and the affective meanings of learning are defined in different timespaces (Schatzki, 2010).

Does practice theory shed light on smart meters?

While the electric current is sourced through sockets and appliances most of the time, energy remains invisible to householders. An argument often raised in the behavioural literature is that through smart meters electricity consumption moves from its invisible status to visible (to individuals). A growing body of literature advocates that in everyday life energy consumption is not a conscious act, but a by-product of performing everyday routines that are powered by energy (Shove & Walkar, 2014). This is acknowledged by Wilhite (2005, p. 2), who notes that 'people do not consume energy per se, but rather things energy makes possible, such as light, clean clothes, travel, refrigeration and so on'. Sociological studies of domestic energy consumption create a route through which energy consumption can be approached indirectly by researching not the energy per se, but the 'enmeshed...network of related practices and habits' (Shove et al., 2012, p. 246). One of the advantages of the practice approach is that it allows understandings of how energy is integrated as a constitutive ingredient for everyday routines (Shove et al., 2012) or in complexes of practice (Schatzki, 2010). For instance, studying migrant families enables paying 'close attention to energy in action and conversely in inaction'. In this conceptualisation different energy-making practices can create distinct 'object-like energies... [that] intersect with other elements to reproduce (or limit) resourcefulness' (Strengers & Maller, 2012, p. 26).

The term 'energy-making practices' refers to the practices of making usable energies, such as chopping wood for a fire or using micro-generation. For example, in some cultures, making energy involves distinct arrangements of 'technical equipment (donkeys, axes, clay pots), practical knowledge, and understandings to produce specific energies that are integrated into practices that are dependent on them (such as cooking)' (Strengers & Maller, 2012, p. 26). The qualities and characteristics of different energy systems define how energies are constituted and become integrated into practice as a distinctive, traceable gathering of material things. A related implication of this is that the supply and demand of energy is no longer viewed as separate and separable, but as a suite of intimately interwoven practices, whereby the technologically mediated energy provision (sockets and wires) interact with the ways energies are consumed and used in the course of everyday life. While studying energy through smart meters as a material of practice could help in understanding the role of energy and its associated technologies in everyday energy demand, energy itself is not material or visible, 'it is not something we touch, listen or smell' (Pink, 2012, p. 69).

Because bodily doings and sayings are classified as part of the material arrangements, practices are by default material. Thus, there is no practice which is not material. Bundling this with Shove et al.'s (2012) conceptualisation of materiality, it is possible to study energy as ingredient of practice – material element that sometimes makes demands on practice through its immateriality or through other tangible technologies (Strengers, 2013). In this framing, energy enables practices to work as being a constitutive element of social practice, as the engagements, know-how, institutional knowledge and technologies. This position extends the understanding of the way energy is handled around the home and how energy becomes a constitutive ingredient of social practices (Walker., 2014). The classification of practice elements in know-how (or embodied habits), engagements, institutional knowledge and technologies is useful in understanding the constitutive role of energy in accomplishing a practice in time and space, and how practices emerge, develop and fossilise (Shove et al., 2012). This view of energy as an ingredient of practice underlines the view 'that individual technologies add value only to the extent that they are assembled together into effective configurations' (Shove et al., 2012, p. 10).

Key concepts to understand smart meter energy demand data: other theoretical traditions

Besides behavioural and social practice theories, several theoretical approaches applied to the energy sphere analyse energy demand and can be considered in the context of smart meters.

Household economics plays a strong role in the discussion on energy demand. Behavioural determinants (such as the timing of use per appliance) are generally classified as decisions requiring little or no financial investment. These counterpoise physical determinants of appliances (e.g. the energy efficiency rate of an appliance), which result in relatively 'fixed' decisions (Mansouri et al., 1996). However, the extent to which demand changes is very limited due to the low willingness of consumers to substitute an appliance use for another (Kasulis et al., 1981).

In the energy economics literature, the energy demand has conventionally been associated with high levels of inelasticity. Changes in price over short and medium periods (e.g. via smart meters) would yield only minor changes in electricity demand (Bernard et al., 2011; Borenstein, 2005; Filippini, 1995; Kamerschen & Porter, 2004; Silk & Joutz, 1997). Two factors underpin the inelasticity of residential electricity demand. First, the flat tariffs which most residential consumers have been exposed to over the last decades prevent the existence of signals regarding the true costs of electricity. An increase in price of 100% typically is expected to trigger reduction of around 20% change in demand (EPRI, 2008). Responses to peak period prices lead to price elasticities of 0.02–0.1 (Faruqui & Sergici, 2009). Because of widespread flat tariffs, customers do not operate in a responsive manner in the retail market. The image of a supermarket without price labels on the products can figuratively explain the lack of price signals in electricity markets. One of the unintended consequences of the lack of information on prices is that consumers might simultaneously opt for sub-optimal choices (Dana, 1999). In the energy sector, simultaneous demand at the time of low supply is extremely problematic, as it forces energy companies to produce expensive and frequently highly carbon-intensive extra generation. Second, at least in the developed world, electricity prices are usually too low to make any impact on a large scale. Electricity prices are typically low enough that electricity payments make up only a small portion of the average household's budget, and consumers likely perceive electricity as a necessity during peak times (Thorsnes et al., 2012).

Empirical studies on residential energy demand have tended to focus on one or several of the human factors either causing or being consequences of energy demand. The physical-technical-economic modelling work which has dominated energy analysis over the past 30 years or so places less emphasis on human occupants and more on building thermodynamics and technology efficiencies (Lutzenhiser, 1993). The technical and price data, along with the modelling techniques habitually used in energy demand research, imply that the material, visual and physical tend to prevail over the variable and correlational (Shipworth, 2013). The preponderant aspiration of existing studies has been to model and forecast household energy consumption rather than describe the structure of energy demand (Suganthi & Samuel, 2011; Swan & Ugursal, 2009). The limitations of the technical-economic approaches have led to interventions which had limited impact in creating embedded energy conservation behaviour change (Shirani, et al., 2013).

Social science understandings of the structure of energy demand tend to emphasise the central role of human activities (people's doings) in shaping how and when energy demand occurs. This implies placing the timing of energy services consumption at the centre of research investigating the relationship between people and energy demand. A distinct disciplinary perspectives epitomising how other social science angles value time in relation to the structure of energy demand comes from time geography. Time budget is seen as a concept delimiting the time available for discretionary activities (Jenkins and O'Leary, 1997).

Applying behavioural theories in the context of smart metering

Smart meters and shifting behaviour

While at the beginning energy demand management was always the responsibility of governments and energy companies, with time this role was delegated and later embedded into energy infrastructures such as smart networks or smart technologies (Strengers, 2013). The role of utility companies was always to supply the amount of electricity demanded by consumers, who continually expect to receive this type of service (Shove et al., 2012). In order to fulfil the increase in demand (or make it more efficient), energy companies developed smart infrastructures, such as smart networks and smart technologies, that manage everyday electricity demand. In this view, the focus is on the use of demand management technologies that attempt to engage householders with their electricity demand.

The main applications of behavioural studies in research related to smart meters and energy demand tend to assess how demand changes based on a set of interventions. This is consistent with the ontological proposition of behavioural research, i.e. to shed light on the net effects of a set of inputs on individuals. Interventions are generally represented as variations in the technologies which are provided, the price which is offered and a variety of non-monetary measures, such as higher level of information and social responsibility. Correspondingly, applied behavioural studies have focused their attention on the acceptability of technological change, response to variations in price and information on energy consumption.

Examples of applications of behavioural science have proliferated following the implementation of smart meters in an attempt to understand the net behavioural effects of smart meters from an end-user perspective. From a policy perspective and as stated in previous chapters, the EU Electricity Directive mandates that all member states shall 'equip' 80% of their consumers with 'intelligent metering systems' by 2020, such that householders have at their disposal their real-time energy consumption data at no extra cost, and that they 'are properly informed of actual electricity consumption and costs frequently enough to enable them to regulate their own energy consumption' (CEC, 2009). Currently, smart meters are used in demand reduction applications, specifically in providing improved information feedback to domestic householders about their energy consumption. Some of the smart meter trials suggest that everyday electricity consumption can be reconfigured with demand response technologies that in some cases engage consumers in managing demand (Foulds et al., 2014).

The increased digitalisation of devices, appliances, meters and control systems in the home poses questions of how consumers respond to such changes. Are consumers adaptable to change? Are there barriers in terms of technology adoption? These questions have been addressed in behavioural studies on technological acceptance. Demand response programs are based on rational actor logic, in that it is assumed that consumers benefit from automation and remote control of appliances by paying less for their consumption. Smart metering in this context is seen as a technical provider of opportunities for automating demand responses.

Recent studies suggest that energy consumers who are using smart meters are worried about their privacy and losing control over their time-dependent daily routines (McKenna, et al., 2013). Indeed, research by Krishnamurti et al. (2012) suggested that the more energy consumers believed that smart meters would let their electricity company control their energy use, the less likely they were to desire one. While external control by utility companies is less desirable, energy consumers find the idea to have the opportunity to remotely control their energy consumption attractive. Cole et al. (2008) suggest that the key element of being comfortable when inside a building is the ability of the user to have control over the environment. Furthermore, they claim that individuals experience comfort in different ways and have different needs, and these may change when individuals gather in groups. These groups may develop collective understandings and practices around comfort. It is suggested that human factors should not be included only in new notions of comfort, but also considered in environmental assessment methods as they will affect buildings' performance (Cole et al., 2008).

Studies have shown that people are not efficient at manually managing their comfort levels – however these might be defined – in comparison with automation technologies and are therefore likely to feel less 'comfortable' when inside buildings (Hinton, 2010). Home automation can act back on everyday routines encouraging householders to act in different ways in response to technological signals. For example, 'A busy parent can be making plans to pick up a child by SMS while cooking dinner. At the same time, automation may be enabling another domestic activity, such as laundering, to be performed in the background. (Strengers, 2013, p. 127)'. Thus, even though home automation technologies may provide householders with more control over their everyday lives, there is no guarantee this will be used to reduce energy consumption. Understanding control is important because it mediates the impact of homes on energy demand (Hargreaves et al., 2013).

Behavioural economists have been studying for a long time the relationship between energy prices and consumption and how variations in tariffs can correspond to structural changes in the timing of demand. Much of the literature focuses on whether the structure of demand is elastic (to changes in price) or not. A specific branch of this literature consists of short-term price elasticity, which studies the behavioural change of individual consumers following a short-term (e.g. intra-day) variation in price.

Time-varying pricing consists of charging higher prices during periods of peak demand to encourage customers to reduce their electricity consumption during those times. A time of use tariff typically sets out several peak and off-peak tariffs, which are charged consistently across every day, with the peak period being the most expensive. For example, the Economy 7 tariff involves two prices: a cheaper night-time electricity tariff, which normally operates from midnight for seven hours at night and a more expensive tariff throughout the rest of the day. In the United Kingdom, this is most effective for those customers who use electric heating.

Chapter 8 further explains how smart meters will provide price information that will encourage consumers to weigh up the costs and benefits of consuming electricity at particular times of the day. Price information is communicated in real-time to many meters at once, and the smart meter itself calculates in real-time the energy consumption bill. The use of smart meters that send price information to consumers has been investigated in pilot studies to determine the benefits of smart metering. The Energy Demand Research Project (EDRP) – a smart-metering trial in the United Kingdom, used about 18,000 smart meters – concluded that the aggregated smart-metering data with smart electricity display meters resulted in higher energy saving (around 3% higher) than smart meters without smart electricity display meters.

Dishwashing and laundry practices responded well to the time-of-use tariff scheme while cooking was a 'non-negotiable' practice in households with children (Strengers, 2013). These findings suggest that time-of-use tariff needs to take into account other structures that are connected with householders' everyday lives.

Dynamic pricing is a tool to be used in reshaping the meaning of electricity, by raising the emphasis of electricity to users with regard to practices-that-consume-electricity or by creating new skills to perform practices-that-consume-electricity in more efficient ways (Strengers et al., 2011). This conceptualisation of price tariffs as a disruptor of everyday life already exists and has many similarities to electricity blackouts, which have long been part of the electricity system and that are well known amongst energy consumers (Strengers et al., 2013). Power cuts are disruptive moments, making the running of normal, electricity-dependent activities impossible (or difficult running on batteries) (Rinkinen, 2013). Other disruptions with knock-on effects on electricity consumption are house moves (Gnoth, 2013), home renovations and changing family roles (Shirani et al., 2013).

One could argue that dynamic pricing is different from blackouts, and other forms of disruption in a number of ways. Its main purpose is to disrupt electricity consumption during predictable periods. Second, while power cuts are considered to be the failures of the electricity systems or failure of those who supply it, the main role of dynamic pricing is to make electricity available at all times at an affordable cost. Demand management technologies which make sounds or change colours according to the price of electricity are important in this case, as they are notifying and communicating the exceptional circumstances that should reconfigure the meaning of the electricity for a period of time (Faruqui & Sergici, 2009). Dynamic pricing is a tool for shifting demand or creating cold spots when the normal everyday practices are temporarily suspended in favour of family time (Southerton, 2003). The terms 'cold spots' and 'hot spots' illustrate that activities are not evenly spread during a day and are constrained by collective and institutional rhythms. In this view, while social loads do not always correlate directly with peak loads, by making electricity matter, social and environmental circumstances can influence and disrupt everyday activities that consume electricity. In the following, we explore how energy feedback as a behaviour tool influence and reconfigure everyday routines.

Behavioural studies and feedback on energy consumption

Real-time energy consumption data from smart meters as a form of immediate feedback is thought to be the main mechanism in engaging consumers to manage their electricity use. In this perspective, smart-metering data is collected and transferred to supplier at half-hourly or more regular time intervals as a combination of parameters such as electricity usage values or time span. The actionable data (such as savings by using less or later) is transferred back to the householders via smart electricity display meters or web portals. The ability of the smart meters for bidirectional communication of data enables accurate billing, the implementation of a dynamic tariff scheme and the remote control of electricity load and devices (Goulden et al., 2014). In this data-based environment, smart metering is not only important but is also necessary to learn how to control devices (Strengers, 2013). Extensive reviews of past research focusing on a range of consumer feedback methods and mechanisms exist, and these were also looked at as part of Chapter 6 (Darby, 2010; Faruqui & Sergici, 2009; Langenheld, 2010;).

Up to today, there is no agreement over the best format, form or frequency of the smart-metering data. For example, with an interactive web page to display household consumption, smart-metering data achieved a 15% reduction in electricity consumption from those households that visited the web page at least once (Vassileva et al., 2012). Others suggests that with time the smart meters stop offering new information and 'fallback' into the background of everyday life (Hargreaves et al., 2013). This loss of interest was documented in a government survey, which found that one in five people reported never looking at their smart meters and its associated technology (DECC, 2015).

Similarly, the growing popularity of health and fitness apps and wearables (such as Fitbit One, Jawbone Up or Nike+ Fuelband) is enabling the collection of a large quantity of actionable data that, in turn, can be shared and compared with other data through a range of associated technologies (such as heart rate monitors). Unlike the data from smart meters, the fitness data feeds back to participants how fit they are, where they can improve and how much more training they need to do. Thus, racing or time trialling is oriented towards acting on data: here the engagements, know-how, institutional knowledge and technologies of calculating and establishing benchmarks are essential to the performance of the fitness practice (Strengers, 2013).

While the numerical data as feedback on energy consumption has an important role to play in energy conservation.

Researchers have also suggested feedback mechanisms on behaviour that are more effective than appliance-specific feedback. For example, Harries et al. (2013) underline the importance of providing feedback in terms of activities rather than kilowatt-hour or pounds. Similarly, a practice-based advice (such as how to save energy while doing the washing) and a user-friendly assistance can help understand 'how much energy each practice uses and derive options for change' (Pullinger et al., 2014, p. 1150).

While there is considerable evidence to suggest that behavioural science tools either independently or in combination with other energy demand strategies are capable of achieving resource savings, these strategies leave unanswered questions about how energy is consumed in the everyday and how savings occur. Behavioural models ignore the changing nature of comfort activities. For example, behavioural models that use consumption data as feedback do not challenge consumers on the way they are using electricity; instead, it engages consumers in a limited number of 'energy saving practices' to 'switch off' unnecessary lights, unplug appliances or, under some scenarios, encourage consumers to make energy efficiency changes such as installing energy-efficient light bulbs or an energy efficient refrigerator. Thus, the behavioural tools' focus is on making, managing and using technology more efficiently rather than investigating the importance of the technology in changing energy demand.

The application of economic principles in energy demand management programs is part of the energy policy approach in the form of decision-making (such as cost–benefit analysis) and understanding of the electricity consumption and choice discussed previously in this chapter. These economic principles inform the design and evaluation of, and justification for, smart meters and smart-metering strategies (Nyborg et al., 2013). However, the application of economic principles in understanding everyday electricity consumption is problematic for several reasons. First, the economic models ignore how practices-that-consume-electricity are learned or shared between consumers: 'Consumption is embedded in the way that different social groups engage in practices and that practices are almost always shared enterprises that are performed in the (co-)presence of others and therefore subject to collective norms of contextualized engagement' (Southerton et al., 2004, p. 34). Second, by suggesting that everyday electricity consumption decisions are always calculated, they ignore the habitual nature of electricity consumption (Warde et al., 2012). Third, they ignore the co-dependent relationship of smart infrastructures, smart technologies and wider systems of provision in everyday electricity consumption (Shove, 2010). But most importantly they ignore the ways in which practices-that-consume-electricity change over time.

Applying social practice theory to timing of energy demand and smart meters

Smart meters measure how electricity demand varies in time. They provide aggregate information about load profiles, which in turn are highly correlated with the timing of people's practices. Social practice theory, timing of activities, load profiles and smart meters cannot be easily put on the same scale. Time use studies often bridge this gap as they rely on quantitative data around people's activities. These can be associated with qualitative narratives of social practices as well as quantitative accounts of how electricity demand varies throughout the day.

Several studies measure, simulate or make use of electricity demand data, for instance, developing demand profiles from time use activities based on time use survey and diary data.

Time use studies

There is a growing literature on time use data and how this can inform load profiles and peak demand. Existing studies on time use and energy consumption tend to be based either on synthetic stochastic models or on measured time use survey data.

The timing of social practices can be traced in various ways, including ethnographic studies, questionnaires and GPS methods with national time use surveys offering the most comprehensive information. Time use surveys collect data through time diaries covering 24 hours. Respondents are asked to fill in diaries for one or two randomly designated days, stating what activities were performed and where they were performed.

Findings from time use surveys are summarised in ten-minute intervals. This way it is possible to know, for instance, how many people are watching TV between 19.20 and 19.30 on a specific weekday. Results from national time use surveys are a statistically significant representation of the timing of social practices in a single country. A higher level of aggregation can describe timing of social practices of a whole continent. The Harmonised European Time Use Survey consists of over 200,000 comparable individuals across 15 countries.

Time use surveys specify the structure of everyday life and consequently the structure of energy demand. They capture the timing of some activities taking place outside the household, e.g. travelling by public transport. Other activities outside the household are difficult to reconcile with the timing of energy demand – for instance, 'being at work' does not link this activity with energy demand. For activities inside the household, the matter is strictly related to occupancy data. Some of the time use data can be associated with the timing of electricity consumption. This mainly depends on how appliance-specific the diary entry is. For instance, for the entry called 'TV and Video watching' it is possible to derive not only the timing of electricity demand, but also, with some approximation about the average efficiency of the TV and video sets, the actual power demand in KW. However, diary entries like 'household work' are not appliance specific, making it difficult to associate the timing of that activity with electricity demand. Smart meter data has been used to connect across time use activities and electricity consumption, often relying on activities as proxies for practices to understand variations in energy demand (Ramírez-Mendiola et al., 2017; Torriti, 2017).

Studies based on time use data consist of a growing body of work which typically relies on national time use surveys to either model electricity load profiles or infer energy-related proxies, such as occupancy. Early work comprises a study by Capasso et al. (1994), who modelled 15-minute period consumption patterns based on appliance and homeowner variables; Wood and Newborough (2003), who used three characteristic groups to explain electricity consumption patterns in the household: "predictable", "moderately predictable" and "unpredictable"; Stokes et al. (2004), who modelled domestic lighting with a stochastic approach, generating load profiles with a resolution of 1 minute from the 30-minute resolution of measured data in 100 households; and a study by Firth et al. (2003), who analysed groups of electrical appliances (continuous and standby, cold appliances and active appliances) in terms of time of the day when they are likely to be switched on.

In the United Kingdom, Richardson et al. (2003) make use of the National Time Use Survey to develop a model which generates occupancy data for UK households. The model consists of probabilistic approach to infer how many other occupants enter or leave the household between time intervals. The model is used by Richardson et al. (2010) and Ramírez-Mendiola et al. (2018) to simulate electricity demand. Other have used time use data to derive flexibility indices (Torriti et al., 2015).

In a similar UK study, Blight et al. (2013), examined the occupant behaviour and its impacts on heating consumption in Passivhaus buildings. Using Richardson et al. (2003) model, the authors developed occupancy, appliance-use and door-opening profiles, based on UK Time Use Survey, which describes time use at a ten-min resolution by 11,600 householders. Their finding suggests that the occupancy patterns are less significant factors to the total heating energy than other factors, such as set point temperature and appliance use.

In Sweden, Widén and Wäckelgård (2010) developed a model simulating households' activities based on time use data. The timing of electricity demand is derived from time use data combined with appliance holdings, ratings and daylight distribution. The same author applied the same model to water heating (Widén et al., 2009a) and lighting (Widén et al., 2009b).

In Ireland, Duffy et al. (2010) applied the same probabilistic modelling to five different dwelling types. They compare the synthetic data generated by the model with metered electricity demand. Their findings show unusual peak loads during the day and night which do not correspond to existing load profiles.

In Spain, López Rodríguez et al. (2013) used the National Time Use Survey to generate activity-specific energy consumption profiles or to cluster consumers based on their states of occupancy. They used the generated profiles to identify appliances that were running during the occupancy.

Aerts et al. (2014) using the Belgian time use data define a three-state probabilistic model to generate occupancy patterns. They combine socio-economic aspects of population with occupancy data in investigating the clustering of different occupancy patterns.

Others also consider socio-economic characteristics (such as age, employment status, income or main activity) to be powerful predictors of occupancy characteristics. For example, Dar et al. (2015) using the Norwegian Time Use Survey investigated the effect of occupant behaviour and family size on the energy demand of a building and the performance of the heating systems. They identify nine occupancy categories based on number of occupants and working hours.

Time use data are representative of average days and are simple representations of the structure of energy demand based on the ordering of social practices and everyday life. The studies typically depend on secondary data (i.e. national time use surveys). Large-scale time use surveys are conducted very infrequently, namely every ten years. This means that often researchers connecting time use data to energy demand need to rely on data which is relatively old.

Making sense of social science theories in relation to smart meters

In this chapter we have seen how the timing of energy demand is inevitably connected with individual behaviours and social practices. Behavioural approaches can explain how individuals react to internal (such as individual attitudes, beliefs, values, personal norms) and external factors (such as regulations, institutional constraints and consumption practices). Examples include how people respond to direct feedback (via smart meters) and change the structure of their load profiles. Instead, social practices provide a consistent framework for understanding the structure of energy demand. For instance, the timing of energy demand of a household is the result of the ordering and sequence of a set of practices performed by the household. Equally, phenomena like peak demand can be understood as the outcome of the high level of synchronisation of what people do at specific times of the day.

This chapter provided a review of the key concepts underpinning both behavioural and social practice key concepts in an attempt to conceptualise the role of smart meters. Applied research on energy demand was reviewed with specific reference to behavioural studies on smart metering and time use studies connecting people's activities with load profiles.

Policy intervention in the area of energy demand is dominated by approaches centred on behavioural science. Examples abound, but two are presented here for their proximity with issues related to smart meters.

First, in 2012 the European Commission's provisions of the Third Energy Package set a target of 80% roll-out of smart meters by 2020. One of the key principles underpinning this unprecedented large-scale European project is that direct feedback can improve end users' awareness as well as their participation in smart grids. The effects of direct feedback are based on psychological theory of feedback, which dates back to the 1930s, with Skinner's (1938) model of operant conditioning: behaviours which produce a positive effect are more likely to be repeated than those which produce a negative effect. According to this view, positive results could be seen as a positive reinforcement and negative results as a punishment, thus respectively encouraging or discouraging subsequent behaviour. In the 1960s, feedback was connected with goals and information about progress towards goals (Bandura, 1969). In the 1990s Feedback Intervention Theory appears and asserts that behaviour is regulated by comparisons to pre-existing goals, as behaviour is generally goal directed, and that people use feedback to evaluate their behaviour in relation to their goals. Both negative and positive feedbacks can bring about behavioural change (Vallacher & Wegner, 1987). These theories suggest that feedback is effective provided that the individual focuses on the feedback goal. This means that feedback will have to direct the attention of the individual to a specific goal compared with the status quo. Availability and visibility feedbacks are two necessary conditions for the success of feedback. At an empirical level, over a hundred empirical studies of energy feedback have been conducted over the past 40 years and over 200 articles have been published about energy feedback during that time (Karlin et al., 2014). What transpires from the behavioural approach on smart meters is that the literature is mixed on the conservation effect associated with smart meters, but the most widely cited range is between 3% and 15%. However, the net conservation effects associated with direct feedback are much less significant than it was thought a decade ago, when these were estimated in the region of 6%–10% (Darby, 2008). Recent advances in smart grid technology have enabled larger sample sizes and more representative sample selection and recruitment methods for smart-metering trials. By analysing these factors using data from current studies, McKerracher and Torriti (2013) show that a realistic, large-scale conservation effect from feedback is in the range of 3%–5%. The diminishing net conservation effect of direct feedback on energy demand suggests that the behavioural approach should be integrated with social practice approaches which disentangle the conundrum of what happens behind the meter and better link energy demand with time. This paper attempts to do so by measuring time dependence of energy-related practices.

Second, in the UK Ofgem has been considering the introduction of dynamic tariffs, following recommendations by the Competition and Markets Authority (2016). Dynamic tariffs are designed to be cost-reflective and include time of use tariffs, which encourage consumers to shift their consumption away from times when demand on the network is higher (peak periods) or reduce it altogether. In other words, dynamic pricing should increase the flexibility of the demand side, thanks to the exposure to a varying price of electricity. The main principle supporting the effectiveness of dynamic tariffs is price elasticity, i.e. the willingness and ability of individual households to respond to price signals by altering energy consumption. Price elasticity estimates for residential electricity demand vary widely across the literature. Alberini & Filippini (2011) review a number of studies and suggest that differences might be due to the sample period, the nature of the data, geography and level of aggregation of the data. Recent empirical studies show that the price elasticity of electricity consumption ranges from as low as −0.06 (Blázquez et al., 2013) to as high as −1.25 (Krishnamurthy & Kriström, 2013). In general, literature reviews in this area seem to suggest that the price elasticity of demand for electricity is low. For instance, a meta-analysis by Espey and Espey (2004) finds that the median short run elasticity for 36 studies is −0.28. So, for intra-day price elasticity (or extreme short run price elasticity) there seems to be evidence that parents, for example, do not postpone by an hour the moment when they drop children to school to take advantage of a profitable off-peak tariff. A recent study by a group of Swedish researchers looks that elasticity changes during the day for lighting and cooking and much less so for heating (Broberg et al., 2015). Hence, the engagement of end users in dynamic tariffs depends mainly on the extent to which everyday life and household routines can be reconciled with the new tariffs (Buryk et al., 2015). Hence, price in isolation does not explain changes in energy demand over time. This paper makes a conscious shift to deploy time and practices, and not price and consumption, as the main unit of analysis to understand the relationship between time and energy demand.

Behavioural interpretations of energy demand as consumption which can be modified, thanks to smart meters acting as direct feedback points, call for alternative approaches in which the status quo and interventions take into account collective practice (Gram-Hanssen, 2013; Lutzenhiser, 1993). This is recognised in institutional efforts to make the most of social sciences and their interpretation of energy demand. For example, DECC states that 'for an intervention to have any impact, organisations need to either invest in new technologies and materials or promote changes in practices' (DECC, 2014, p. 24).

Whilst behavioural approaches have seen large deployment in policy intervention and utility-level trial designs, the same cannot be said about practice theories. For example, in recent years, behavioural approaches have been institutionalised through behavioural insights, units and 'nudge' teams in government departments, especially in developed English-speaking countries. Both conceptual and methodological challenges explain why the usability of practice theory research in the policy realm is limited. In this context, quantitative accounts of the timing of practices and how these relate to energy demand are generally missing. The expanding area of empirical time use studies on energy demand will probably fill this research gap in the forthcoming years, also thanks to digital innovation which provides opportunities to analyse big data on everyday life which are more accurate and more frequent than national time use surveys. This chapter showed that this is possible and that smart meters could play a vital role not only in the study on individuals' behavioural change following information and price inputs, but also how bundles of practices vary across time and space.

References

Aerts, D., Minnen, J., Glorieux, I., Wouters, I., & Descamps, F. (2014). A method for the identification and modelling of realistic domestic occupancy sequences for building energy demand simulations and peer comparison. Building and. Environment, 75, 67–78.

Alberini, A., & Filippini, M. (2011). Response of residential electricity demand to price: The effect of measurement error. *Energy Economics, 33*(5), 889–895.

Bandura, A. (1969). *Principles of behaviour modification*. New York: Hold, Rinehart & Winston.

Bernard, J. T., Bolduc, D., & Yameogo, N. D., (2011). A pseudo-panel data model of household electricity demand. *Resource and Energy Economics, 33*, 315–325.

Blázquez, L., Boogen, N., & Filippini, M. (2013). Residential electricity demand in Spain: New empirical evidence using aggregate data. *Energy Economics, 36*, 648–657.

Blight, T.S., Coley, D.A. (2013). Sensitivity analysis of the effect of occupant behaviour on the energy consumption of passive house dwellings. *Energy and Buildings. 66*, 183–192.

Borenstein, S. 2005. The long-run efficiency of real-time electricity pricing. *Energy Journal, 26*, 93–116.

Broberg, T., Brännlund, R., Kazukauskas, A., Persson, L., & Vesterberg, M. (2015). An electricity market in transition demand flexibility and preference heterogeneity. [Accessed on 1 May 2016] Retrieved from https://www.ei.se/Documents/ Publikationer/rapporter_och_pm/Rapporter%202015/Rapport_An_electricity_ market_in_transition_Umea_universitet.pdf

Burningham, K., Venn, S., Christie, I., Jackson, T., & Gatersleben, B. (2014). New motherhood: A moment of change in everyday shopping practices? *Young Consumers, 15*(3), 211–226.

Buryk, S., Mead, D., Mourato, S., & Torriti, J. (2015). Investigating preferences for dynamic electricity tariffs: The effect of environmental and system benefit disclosure. *Energy Policy, 80*, 190–195.

Capasso, A., Grattieri, W., Lamedica, R., & Prudenzi, A. (1994). A bottom-up approach to residential load modeling. *IEEE Transactions on Power Systems, 9*(2), 957–964.

CEC (2009). Directive 2009/72/EC of the European Parliament and of the Council of 13 July 2009 concerning common rules for the internal market in electricity and repealing Directive 2003/54/EC. Retrieved from http://eurlex.europa.eu/ LexUriServ/LexUriServ.do?uri=OJ:L:2009:211:0055:0093:EN:PDF⟩

Cole, R.J., Robinson, J., Brown, Z., O'Shea, M. (2008). Re-contextualizing the notion of comfort. Building Research and Information, *36*, 323–336.

Competition Market Authority. (2016). Provisional decision on remedies. https:// assets.digital.cabinet-office.gov.uk/media/56efe79040f0b60385000016/EMI_ provisional_decision_on_remedies.pdf

Dana J. D. (1999). Using yield management to shift demand when the peak time is unknown. *The Rand Journal of Economics, 30*(3), 456–474.

Dar, U. I., Georges, L., Sartori, I., Novakovic, V. (2015). Influence of occupant's behavior on heating needs and energy system performance: a case of well-insulated detached house in cold climates. *Building Simulation, 8*, 499–513.

Darby, S. (2008). Energy feedback in buildings - Improving the infrastructure for demand reduction. *Building Research and Information, 36*, 499–508.

Darby, S. (2010). Smart metering: What potential for householder engagement? *Building Research and Information 38*(5), 442–457.

DECC. (2014). Developing DECC's evidence base. [Accessed on 1 May 2016]. Retrieved from https://www.gov.uk/government/uploads/system/uploads/attachment_data/file/270126/FINALDeveloping_DECCs_Evidence_Base.pdf

Duffy, A., McLoughlin, F., & Conlon, M. (2010). The generation of domestic electricity load profiles through Markov chain modelling. *3rd International Scientific Conference on Energy and Climate Change* 7–8 October 2010 Athens, Greece.

Energy Saving Trust. (2011). *Trigger points – A convenient truth: Promoting energy efficiency in the home.* London: Energy Saving Trust.

EPRI. (2008) Price Elasticity of demand for electricity: A primer and synthesis, electric electricity research institute, January. Retrieved from /http://my.epri.com/portal/server.pt?Abstract_id=000000000001016264S.

Espey, J. A., & Espey, M. (2004). Turning on the lights: A meta-analysis of residential electricity demand elasticities. *Journal of Agricultural and Applied Economics, 36*, 65–81.

Faruqui, A., & Sergici, S. (2009). Hoshehold response to dynamic pricing of electricity—a survey of experimental evidence. The Brattle Group. Retrieved from /http://www.science.smith.edu/jcardell/Readings/uGrid/House%20Demand Resp%20Experience.pdfS.

Filippini, M. (1995). Swiss residential demand for electricity by time-of-use. *Resource and Energy Economics, 17*, 281–290.

Fine, G. A. (1990). Organizational time: Temporal demands and the experience of work in restaurant kitchens. *Social Forces, 69*(1), 95–114.

Firth, S., Lomas, K., Wright, A., & Wall, R., (2003). Identifying trends in the use of domestic appliances from household electricity consumption measurements. *Energy and Buildings, 40*, 926–936.

Foulds, C., Royston, S., Buchanan, K., & Hargreaves, T. (2014). The many faces of feedback: Beyond the kWh. In C. Foulds & C. L. Jensen (eds.), *Practices, the built environment and sustainability – A thinking note collection* (pp. 19–21). Cambridge. https://www.forskningsdatabasen.dk/en/catalog/2398066223

Foulds, C., Powell, J., & Seyfang, G. (2015). How moving home influences appliance ownership: A Passivhaus case study. *Energy Efficiency, 9*, 455–472

Gnoth, D. (2013). Moving home and changing behaviour – Implications for increasing household energy efficiency. *ECEEE Summer Study Proceedings: Rethink, Renew, Restart.* France: Belambra Presqu'île de Giens

Goulden, M., Bedwell, B., Rennick-Egglestone, S., Rodden, T., Spence, A. (2014). Smart grids, smart users? The role of the user in demand side management. *Energy Research and Social Science, 2*, 21–19.

Gram-Hanssen, K. (2013). Efficient technologies or user behaviour, which is the more important when reducing households' energy consumption? *Energy Efficiency, 6*(3), 447–457.

Guerin, D. A., Yust, B. L., Coopet, J. G. (2000). Occupant predictors of household energy behavior and consumption change as found in energy studies since 1975. *Family and Consumer Sciences Research Journal, 29*, 48–80.

Hargreaves, T., Nye, M., & Burgess, J. (2010). Making energy visible: A qualitative field study. *Energy Policy, 38*(10), 6111–6119.

Hargreaves, T., Nye, M., Burgess, J. (2013). Keeping energy visible? Exploring how householders interact with feedback from smart energy monitors in the longer term. *Energy Policy*, 52, 126–134.

Harries, T., Rettie, R., Studley, M., Burchell, K., Chambers, S. (2013). Is social norms marketing effective? A case study in domestic electricity consumption. *European Journal of Marketing, 47*(9):1458–1475.

Hinton, E. (2010). Review of the literature relating to comfort practices and socio-technical systems. Paper 35, Environment, Politics and Development Working Paper Series, London: Department of Geography, King's College London.

Jenkins, S. P., & O'Leary, N. C. (1997). Gender differentials in domestic work, market work, and total work time: UK time budget survey evidence for 1974/5 and 1987. *Scottish Journal of Political Economy, 44*(2), 153–164.

Kamerschen, D. R., & Porter, D. V., 2004. The demand for residential, industrial, and total electricity, (1973–1993). *Energy Economics, 26*, 87–100.

Karlin, B., Ford, R., Squiers, C. (2014). Energy feedback technology: A review and taxonomy of products and platforms. *Energy Efficiency, 7*, 377–399.

Kasulis, J. J., Huettner, D. A., & Dikeman, N. J. (1981). The feasibility of changing electricity consumption patterns. *Journal of Consumer Research, 83*, 279–290.

Krishnamurti, T., Schwartz, D., Davis, A., Fischhoff, B., de Bruin, W. B., Lave, L., & Wang, J. (2012). Preparing for smart grid technologies: A behavioral decision research approach to understanding consumer expectations about smart meters. *Energy Policy, 41*, 790–797.

Langenheld, A. (2010). Advanced metering and consumer feedback to deliver energy savings – Potentials, Member States experience and recommendations Report prepared by the Joint Research Centre of the European Commission: Ispra.

López-Rodríguez, M.A., Santiago, I., Trillo-Montero, D., Torriti, J., Moreno-Munoz, A. (2013). Analysis and modeling of active occupancy of the residential sector in Spain: an indicator of residential electricity consumption. *Energy Policy, 62*, 742–751.

Lutzenhiser, L. 1993. Beyond price and attitude: Energy use as a social process. *Annual Review of Energy and the Environment, 18*, 259–263.

Mansouri, I., Newborough, M., & Probert, D. (1996). Energy consumption in UK households: Impact of domestic electrical appliances. *Applied Energy, 54*, 211–285.

McKenna, E., Richardson, I., & Thomson, M. (2012). Smart meter data: Balancing consumer privacy concerns with legitimate applications. *Energy Policy, 41*, 807–814.

McKerracher, C., & Torriti, J. (2013). Energy consumption feedback in perspective: Integrating Australian data to meta-analyses on in home displays. *Energy Efficiency, 6*(2). 387–405.

Nicolini, D. (2012). *Practice theory, work, and organization – An introduction*. Oxford: Oxford University Press.

Nyborg, S., Røpke, I. (2013). Constructing users in the smart grid - insights from the Danish eFlex project. *Energy Efficiency, 6*, 655–670.

Pink, S. (2012). *Situating everyday life: Practices and places*. London: Sage.

Pullinger, M., Heather L., Webb J. (2014), Influencing household energy practices: a critical review of UK smart metering standards and commercial feedback devices. *Technical Analysis and Strategic Management, 26,* 1144–1162.

Ramírez-Mendiola, J. L., Grünewald, P., & Eyre, N. (2017). The diversity of residential electricity demand–a comparative analysis of metered and simulated data. *Energy and Buildings, 151,* 121–131.

Ramírez-Mendiola, J. L., Grünewald, P., & Eyre, N. (2018). Linking intra-day variations in residential electricity demand loads to consumers' activities: What's missing? *Energy and Buildings, 161,* 63–71.

Reckwitz, A. (2002). Toward a theory of social practices: A development in culturalist theorizing. *European Journal of Social Theory, 5*(2), 243–263.

Richardson, I. W., Thomson, A. M., & Infield, D., (2003). A high-resolution domestic building occupancy model for energy demand simulations. *Energy and Buildings, 40,* 1560–1566.

Richardson, I. W., Thomson, A. M., Infield, D., & Clifford, C., (2010). Domestic electricity use: A high-resolution energy demand model. *Energy and Buildings, 42,* 1878–1887.

Rinkinen, J. (2013). Electricity blackouts and hybrid systems of provision: Users and the 'reflective Practice. *Energy Sustainability and Society, 3*(1),1–10.

Røpke, I. (2009). Theories of practice – New inspiration for ecological economic studies on consumption. *Ecological Economics, 68*(10), 2490–2497.

Schatzki, T. (1996). *Social practices: A Wittgensteinian approach to human activity and the social.* New York and Cambridge: Cambridge University Press.

Schatzki, T. (2009). Materiality and social life. *Nature and Culture, 5*(2), 123–149.

Schatzki, T. R. (2010). *The timespace of human activity: On performance, society, and history as indeterminate teleological events.* Washington DC: Lexington Books.

Shipworth, D. (2013). The vernacular architecture of household energy models. Perspectives on Science, 21(2), 250–266.

Shirani, F., Butler, C., Henwood, K., Parkhill, K., Pidgeon, N. (2013). Disconnected Futures: Exploring Notions of Ethical Responsibility in Energy Practices. *Local Environment, 18,* 455–468.

Shove, E. (2003). *Comfort, cleanliness and convenience.* Oxford: Berg.

Shove, E. (2010). Beyond the ABC: Climate change policy and theories of social change. *Environment and Planning A, 42*(6), 1273–1285.

Shove, E., Trentmann, F., & Wilk, R. (2009). *Time, consumption and everyday life: Practice, materiality and Culture.* Oxford: Berg.

Shove, E., Pantzar, M., & Watson, M. (2012). *The dynamics of social practice: Everyday life and how it changes.* London: Sage Publications.

Shove, E., & Walker, G. (2014). What is energy for? Social practice and energy demand. *Theory, Culture & Society, 31*(5), 41–58.

Shove, E. and Warde, A. (2002). Inconspicuous consumption: the sociology of consumption, lifestyles and environment. In R. Dunlap (Ed.), Sociological theory and the environment: classical foundations, contemporary insights. (pp. 230–241). Lanham: Rownan and Littlefield Publishers.

Shirani, F., Butler, C., Henwood, K., Parkhill, K., & Pidgeon, N. (2013). Disconnected futures: Exploring notions of ethical responsibility in energy practices. *Local Environment, 18*(4): 455–468.

Silk, J. I., & Joutz, F. L., (1997). Short and long-run elasticities in US residential electricity demand: A co-integration approach. *Energy Economics, 19,* 493–513.

Sintov, N. D., Dux, E., Tran, A., & Orosz, M. (2016). What goes on behind closed doors? How college dormitory residents change to save energy during a competition-based energy reduction intervention. *International Journal of Sustainability in Higher Education, 17*(4), 451–470.

Skinner, B. F. (1938). *The behaviour of Organisms: An experimental analysis.* New York: Appleton-Century.

Southerton, D. (2003). Squeezing time: Allocating practices, coordinating networks and scheduling society. *Time & Society, 12*(1), 5–25.

Southerton D. (2006). Analysing the temporal organization of daily life: Social constraints, *Practices and their Allocation. Sociology, 40*(3), 435–454.

Southerton, D, Warde, A., & Hand, M. (2004). The limited autonomy of the consumer: Implications for sustainable consumption. In D. Southerton., B. Van Vliet., & H. Chappells (eds.), *Sustainable consumption: The implications of changing infrastructures of provision* (pp. 32–48). Edward Elgar, Cheltenham.

Stern P. C. (2000). Toward a Coherent Theory of Environmentally Significant Behavior. Journal of Social Issues, 56, 407–424.

Stokes, M., Rylatt, M., & Lomas, K. (2004). A simple model of domestic lighting demand. *Energy and Buildings, 36*, 103–116.

Strengers, Y., Maller. C. (2012). Materialising energy and water resources in everyday practices: Insights for securing supply systems. *Global Environmental Change, 22*, 754–763.

Strengers, Y. (2013). Smart Energy Technologies in Everyday Life: Smart Utopia? Palgrave Macmillan, London (2013).

Strengers, Y. (2011). Negotiating everyday life: The role of energy and water consumption feedback. Journal of Consumer Culture, 11, 319–338

Suganthi, L., & Samuel, A. A. (2011). Energy models for demand forecasting—A review. *Renewable and Sustainable Energy Reviews, 16*, 1223–1240.

Swan, L. G., & Ugursal, V. I. (2009). Modeling of end-use energy consumption in the residential sector: A review of modeling techniques. *Renewable and sustainable energy reviews, 13*(8), 1819–1835.

Thorsnes, P., Williams, J., & Lawson, R. (2012). Consumer responses to time varying prices for electricity. *Energy Policy, 49*, 552–561.

Torriti, J. (2017). Understanding the timing of energy demand through time use data: Time of the day dependence of social practices., 25, 37–47.

Torriti, J., Hanna, R., Anderson, B., Yeboah, G., & Druckman, A. (2015). Peak residential electricity demand and social practices: Deriving flexibility and greenhouse gas intensities from time use and locational data. *Indoor and Built Environment, 24*(7), 891–912.Vallacher, R., & Wegner, D. (1987). What do people think they are doing? The presentation of self through action identification. *Pyschological Review, 94*(1), 3–15.

Vassileva, J. (2012). Motivating participation in social computing applications: a user modeling perspective. *User Modeling and User-Adapted Interaction, 22*, 177–201.

Vihalemm, T., Keller, M., & Kiisel, M. (2015). *From intervention to social change. A guide to reshaping everyday practices.* London: Ashgate.

Walker, G. (2014). The dynamics of energy demand: change, rhythm and synchronicity. *Energy Research and Social Science, 1*, 49–55.

Warde, A. (2005). Consumption and theories of practice. *Journal of Consumer Culture, 5*(2), 131–153.

Warde, A. (2013). What sort of a practice is eating? In Sustainable Practices: Social theory and climate change, Shove E, Spurling N (eds.). Abingdon: Routledge. 17–30.

Widén, J., Lundh, M., Vassileva, I., Dahlquist, E., Ellegård, K., & Wäckelgård, E. (2009a). Constructing load profiles for household electricity and hot water from time-use data—Modelling approach and validation. *Energy and Buildings, 41*(7), 753–763.

Widén, J., Nilsson, A. M., & Wäckelgård, E. (2009b). A combined Markov chain and bottom-up approach to modelling of domestic lighting demand. *Energy and Buildings, 41*(10), 1001–1012.

Widén, J., & Wäckelgård, E. (2010). A high-resolution stochastic model of domestic activity patterns and electricity demand. *Applied Energy, 87,* 1880–1892.

Wilhite, H. (2005). Why energy needs anthropology. *Anthropology Today, 21,* 1–2.

Wilhite, H. (2012). Towards a better accounting of the roles of body, things and habits in consumption. In: A. Warde & D. Southerton, (eds.), *The habits of consumption* (pp. 87–99). Helsinki: COLLeGIUM of Studies Across Disciplines in the Humanities and Social Sciences.

Wilhite, H., Shove, E., Lutzenheiser, L., & Kempton, W. (2000). The legacy of twenty years of energy demand management: We know more about individual behaviour but next to nothing about demand. In Jochem, E. Sathaye, J. Bouille, D. (eds), *Society, behaviour and climate change mitigation* (pp. 109–126). Dordrecht: Kluwer Academic Publishers.

Wood, G., & Newborough, M. (2003). Dynamic energy-consumption indicators for domestic appliances: Environment, behaviour and design. *Energy and Buildings, 35,* 821–841.

8   A smart future?

Is the energy system changing?

Energy systems in most developed countries are experiencing historic change. For the first time since the Industrial Revolution, the United Kingdom had a full day where no coal was used to produce electricity on 21 April 2017. This has been repeated on so many full days, since that the absence of coal in the energy supply mix has become a new normality.

The UK government's support for renewables is arguably a success story, with £2.5 billion invested to support low carbon innovation between 2015 and 2021. The decarbonisation of the system is one of the standout energy successes of the United Kingdom over the last couple of decades. UK emissions in 2016 were 42% lower than 1990 (BEIS, 2017a). Demand played a vital role in decarbonisation. The shift towards a service-led economy has been paralleled with the decline of manufacturing and heavy industries. Between 1990 and 2016 industrial electricity consumption fell from 38.7 million tonnes of oil to 23.7 million tonnes (BEIS, 2017b). Despite their impact and much progress on decarbonisation, the United Kingdom still has a long way to go if it is to meet its legally mandated target to cut emissions by 57 per cent by 2032 compared with 1990 levels. Low-carbon energy would need to reach roughly three-quarters of UK power generation for this goal to be achieved.

From a policy perspective, the Energy Trilemma has been at the centre of the UK energy system since the previous significant Bill reforming the energy system (the Electricity Market Reform in 2008). The Trilemma places much emphasis on security of supply. For gas, security of supply consists of ensuring that geopolitical tensions do not bring about shortages. For electricity, security of supply is often accompanied by misconceptions around 'keeping the lights on'. For example, a House of Lords report in 2016 raised unnecessary alarms over security of supply during the Christmas festive season. Concerns over security of supply included asking why baseload power was prioritised at the expense of flexible infrastructure.

Like in other markets, where sudden increases in demand push prices up, for decades energy utilities have been making money at times of peak demand, thanks to higher marginal prices. In several countries, in order to increase the penetration of renewables, legal incentives have been set in place so that System Operators prioritise renewables over other forms of generation. This means that renewables, whose generation is intermittent as it depends on weather, have become baseload providers. This has decreased the costs of generating electricity for baseloads, i.e. non-peak levels of electricity demand. In the past, when utilities were vertically integrated companies, they ran generation at baseload power during off-peak periods, while at peak times they added to the supply mix plants which could be easily powered up and down. In recent years, the unbundling of vertically integrated companies and deregulation of the industry implies that generation needs to take into account the marginal cost of electricity. This means that having stand-by plants to run for only a few days a year when demand is higher has become less and less cost-effective.

The popularity of renewable energy is higher than ever with 79% of the public in support. This level of support might be even higher should the full price impacts of renewables on electricity bills become a reality. For instance, prices for electricity in Germany dropped below zero more than 100 times in 2017 alone, meaning that (mostly large industrial) customers are being paid to consume power.

Costs have been plummeting too, with solar PV and battery being increasingly cost-effective and delivering at a significantly large scale. For instance, in 2017 GB achieved a record of 8.7 GW of solar production whilst from a global perspective capital costs of solar have been decreasing more than expected. About 73 GW of new solar PV were built globally in 2016, and in recent years PV production has increased by 50% each year compared with the previous year. These developments have also been based on growing research investment. For example, the Research Councils UK Energy Programme invested £839 million between 2003 and 2011, and a further £625 million were invested between 2012 and 2018. Much of the research and technological enthusiasm around smart grids has brought about increasing emphasis on the role of smart meters. Smart-metering advocates comprise a range of professionals, such as industry analysts, technology developers and academic researchers. According to their views, the economic benefits of smart meters are often presented without much critical insight, especially in media articles and marketing materials.

Moving beyond factual policy and technological change, conceptual interpretations of this transition involve systems-in-systems visions with a complexity due to interwovenness or interrelatedness of actors, perspectives, dimensions, factors, etc. Because the system involves several interlinked systems at different levels, these all impact on each other. According to this view, complexity and interrelatedness are not embedded in policy and market instruments whose objective is to foster innovation. The lack of systems-in-systems approaches means that typically support goes to single products and technologies, not system integration, nor to support the change processes (Geels, 2004). This poses questions around which configurations will take place in the future and the role smart meters will have in such context.

Smart meters and the automated future

Like other technologically induced change, smart meters are the meeting point of expectations around improved efficiency and increased interoperability of smart demand on one side and practices, behaviours and social change broadly defined on the other side.

Smart meters, along with other smart home technologies (fridges, thermostats, digital voice assistants and phones), are often seen as key features of a more automated, technology-driven future.

In policy discourses smart meters, grids and homes are often described as 'revolutionary' (European Commission, 2006) and 'disruptive' (World Economic Forum, 2017). It is repeatedly assumed that they can significantly change and improve the future (Strengers & Nichols, 2017). This transformation will involve automating electricity demand, presumably through distributed forms of generation, higher electrification and providing end users with real-time demand data which will empower the 'smart' consumer to proactively engage with the balancing of demand and supply.

In general interoperability, two-way communications, big data and Internet of Things are associated with the term 'smart' (Van Dam et al., 2010). Darby (2018) provides different definitions of what 'smart' actually means. In this context, smart meters are seen not only as intermediaries, in the sense that they operate between men and women and appliances, but also as intermediate devices, because they stand between a past of human labour, manual contact with household activities and a future made of automation and control.

Representations of higher levels of automation abound both in fiction and in the 'smart'-related literature. One of the narratives evolves around the vision of algorithm-enabled controllers which are delegated the task of turning appliances and devices on or off based on a range of sensors, remote applications and inputs. Thus, an alternative to active monitoring of consumption via smart meters consists of automated demand controllers, which control the load according to pre-set parameters, e.g. occupancy patterns, ambient temperature outputs, average seasonal external temperature, comfort indexes. Demand controllers vary in type, from Computer-Aided Home Energy Management (Williams et al., 2006) and Wireless Sensor Network applied to electrical power management (Molina-Garcia et al., 2007) to multi-level optimisation mechanisms for customer-side load management and multi-agent home automation system for power management. The automated demand controller enables floor level balancing of electricity demand and supply. It collects a number of demand inputs from the electricity services and appliances. It then classifies the demand inputs into groups according to their priority level. The low priority group consists of highly controllable demand inputs. The medium priority group entails medium controllability of demand inputs, whereas the high priority group is made of inputs which have low controllability. The prioritisation of groups determines the overall level of demand. When the total electricity demand is greater than the available total electricity supply, the automated demand controller will take into account priority groups based on demand inputs. The demand control will occur from the high priority group to the low priority group based on the decreasing order of controllability of demand inputs. Automated demand controllers take stock of all the demand inputs, even those from medium priority and high priority groups, to measure the size of the new total demand and assess this with the size of the available supply. When the total demand of all groups exceeds the total supply, the automated demand controller will change priority by moving to the next group and repeat the operation. When the total demand level is lower than the supply level, the automated demand controller will stop implementing control actions and move to the next control step. When total demand is less than total supply, a load recovery component is initiated, which is designated to recover the demand high priority group. The automated demand controller measures the new total demand against the total supply based on the total demand including the demand associated with the recovery. If the total supply still exceeds the new total demand, the automated demand controller moves to the next priority group. This operation is repeated until the total supply is lower than the total demand. Overall, load controllers do not bring about reductions in energy demand, but they tend to flatten the load profile (Torriti, 2014). This is particularly interesting in the light of the integration of new demand services, including heat pumps and electric vehicles, which are likely to have significant impact on load balancing and peak management in buildings. There are various ways in which smart meters are expected to shape and interact with a highly automated future. The different scales of these interactions range from smart homes to smart grids, as it is explained in the following sections.

Smart meters in smarter homes?

Futuristic aspects of smart homes involve devices which perceive householders' emotions, actions and behavioural habits and can make some intelligent actions based on these perceptions. Smart homes depend on information and communication networks to connect appliances and systems to one another and enable remote access or control of a variety of home services (Darby, 2018). The operation of smart home appliances depends on the application and processing of modern technologies, such as the Internet of Things, the Internet and electronic chips. In so doing, smart home appliances claim to achieve anthropomorphic intelligence (Hargreaves & Wilson, 2017). Products can capture and process information through sensors and control chips, and householders can personalise according to their own habits. Smart homes comprise specialised devices that are programmed by a central system. They can communicate with power supply and water source and even automatically turn off the power and water source according to the situation; the smart refrigerator can be defrosted according to the needs of householders; the microwave oven can set the cooking timer according to the scanning strip on the reading food; the smart cabinet can also record the expiration date of the food and provide recipes to residents based on stored food; smart windows can close themselves when it is raining. Smart home devices include televisions, air conditioners, refrigerators, washing machines, door locks, IP cameras, LED bulbs, ovens and smart home sensors, including devices that monitor indoor environments such as smoke detectors, temperature detectors and humidity detectors. These devices and sensors are monitored and controlled in real time by occupants through ubiquitous client devices such as smartphones and tablets.

According to this vision, smart technologies should make energy saving easy; should be easy to use; and should make it possible to do more with less. However, these advantages do not materialise with the immediacy proposed by technology enthusiasts.

> Narratives about smart homes mask and hide the multiple forms of energy consumption that such technologies enable, whilst simultaneously legitimisating and naturalising always-on and constantly updating devices and software. While there are benefits (in efficiency, automation and demand management) we need to consider the 'full story' in order to appreciate the opportunities and problems involved in turning to the smart home as a means of lowering energy demand.
>
> (Hazas & Strengers, 2019, 87)

According to a study by Hargreaves et al. (2018), smart homes require forms of adaptation and familiarisation from householders that can limit how easy they are to use. Smart home technologies may not generate any significant energy savings. On the contrary, there is a risk that consumption increases compared with traditional homes as they facilitate new energy consuming practices, for instance, in relation to escalating expectations about comfort (Nyborg & Røpke, 2011). In this context, smart meters can be connected with various components of the smart home or be bypassed where more interactive communication systems between the smart home and the user become available.

Smart meters and the smart grid

The smart grid is seen as the modern power network infrastructure. It integrates the use of sensors, communications, computing power and control to enhance the overall functionality of the power system (Gellings, 2009). Smart grids are designed to use modern technology to improve operations, maintenance and planning to better management of energy use and costs. Communication, computer, automation and other technologies have been widely applied in the power grid and integrated with traditional power technology. According to this view of smart grids, the role of energy management systems and smart meters is to improve energy efficiency by enabling to move electricity demand of electrical appliances to a time when electricity is available. For example, smart controllers can shift the use time of the smart washing machine to a sunny period, and the energy of the electric power comes from the solar panel (Hanna et al., 2018). This means, for instance, that the use of washing machines is automatically transferred to a period of sufficient power supply, which may reduce their electricity usage. Accurate calculations of real-time consumption and shifting grid loads are designed so that energy consumption is adapted to power supply, rather than adapting actual supply to demand (Paetz et al., 2012). Similarly, one of the advantages of smart homes is that they can improve system efficiency by reducing demand for peak power and based on real-time demand–supply relationships (Gram-Hanssen & Darby, 2018). The relationship between smart grids and smart meters strengthens views of the latter as part of an architecture whose attributes and design functions are determined by the former. This is equivalent to saying that smart meters are a component of infrastructure which is thought more significantly in terms of supply than demand (Danieli, 2018).

Energy as a service

Energy as a service is seen as a possible development in the economics of energy demand, where customers do not get charged for the volumes of energy they use, focusing instead on how they use energy. Rather than paying for the number of kilowatt-hours they consume, consumers will select a tariff such as 'High Comfort'. They could pay a monthly subscription at a competitive fixed price rather than a cost per kilowatt hour.

Like other future developments in this area, energy as a service sets out to make energy savings happen. The service provider is charged with the responsibility of financing, owning, installing and managing the performance of an energy asset from the customer. Prior to this, the service provider carries out an energy assessment in which potential savings opportunities are shared with the end user.

Energy as a service has a wide scope and can range from microgrids, community virtual power plants, community sustainable districts, light as a service, PV as a service, demand response and flexibility services, to heat as a service. Various efficiency-as-a-service models focus on lighting, equipment, software and general energy management. Common solutions include lighting retrofits, upgrades to HVAC and other equipment, building automation and controls, energy storage and water efficiency measures.

In principle energy as a service could reduce operational and performance risks by eliminating customer ownership of equipment, guaranteeing energy savings and tying service payments to performance (Department of Energy, 2008). Even with these benefits, customer inertia, building constraints, utility programme design and lack of stakeholder awareness pose challenges to procuring energy as a service (ACEEE, 2019). The smart meter plays a central role with regard to measurement and verification analysis.

The extent to which smart meter data will be passed on to third parties will determine the boundaries of energy as a service. More broadly, third party access may bring about an evidence base able to serve wider societal goals at national and local levels by enabling and supporting improvements in research, analysis, prediction, evaluation and targeting of public policy (Sustainability First and Centre for Sustainable Energy, 2019). The example which is often cited consists of the Green Button initiative in the United States, which enables data sharing with regard to energy demand. The common format of the smart-metering data facilitates third party access at a low cost. In essence, third party access might have wider functions and applications (also in relation to non-energy activities) than a narrow definition of energy as a service.

Blockchains

A blockchain network technology is a distributed and decentralised database, which records all the transactions transpired between two parties (Crosby et al., 2016). Blockchain technology has recently drawn interest in the energy sector from supply firms, start-ups, technology developers and financial institutions. In the wholesale energy market, trading might be performed through blockchains because of its complex procedures including third party intermediaries, brokers, trading agents, logistics providers, banks and the regulators, which currently add costs to transactions. Blockchains make the simple proposition of removing these intermediaries, hence allowing direct trade with consumers. The blockchain provides a platform that could facilitate the trading between different parties to improve the audibility and process integrity. The scope of the application of blockchains ranges from decentralised power generation, distributed grid management, smart metering and smart communication without relying on a third party (IRENA, 2018). In principle, blockchains could also support more complex functionalities like consumption-based purchasing, selective purchasing and monitoring carbon impact.

In combination with smart meters, blockchains could enable the transition to automated billing for consumers and distributed generators. This is because blockchains could trace energy produced and consumed at each smart meter point. For instance, blockchains could inform consumers about the origins and cost of their energy supply, making energy charges more transparent (Andoni et al., 2019). Higher traceability is critical in order to avoid gaming and benchmarking issues in demand response, especially in the residential sector, where the costs of monitoring would be too high should other technology need to be applied (Torriti, 2017).

An application which was trialled in the United States consists of smart meters which transform energy in equivalent tokens that can be traded in the local marketplace. For instance, these tokens could indicate that a certain amount of energy is produced from the solar panels and can be transferred from a prosumer's smart meter to end-consumers by use of blockchains. Tokens are deleted by the consumer's smart-metering device, as purchased energy is used in the house (Mengelkamp et al., 2018). In the UK trials are being rolled out under the framework of the energy regulator's regulatory sandbox, enabling companies to test new energy business models without being subject to all of the standard regulatory requirements (Fell et al., 2019).

Significant uncertainty surrounds the actual adoption of blockchains in combination with smart meters as standards would need to be defined to allow interoperability between these two technologies. Peer-to-peer energy trading is currently not allowed under law in the vast majority of countries around the world. In addition, once a blockchain system is deployed, any changes in the ruling protocols or codes need to be approved by the system nodes (Andoni et al., 2019). A 2018 study finds that people are more likely to state they would participate in blockchains when the scheme is framed as being run by their local council, followed by an energy supplier, community energy organisation, and social media company. This reduces the range of applications as blockchain may prove insufficiently agile in its deployment in several domains, such as energy company operations, wholesale energy trading and supply, imbalance settlement, digitalisation and decentralised energy.

Smart meters and future changes to tariff approaches

As mentioned in previous chapters, smart meters comprise a set of functionalities which can enable tariffs which are not available with traditional meters: volumetric Time of Use (ToU) tariffs, tariffs based on actual capacity, tariffs based on agreed capacity, real-time pricing, critical peak pricing, critical peak rebates and block pricing. These are described in this section along with a high-level discussion on their respective advantages and disadvantages.

Volumetric time of use tariffs

Under volumetric ToU tariffs, customers are charged for their actual consumption during different periods of the day. Different unit rates can be assigned to set periods of the day called 'time bands', which are defined by the distribution network companies (DNOs) in advance of the charging year and reflect the probability that the network will be congested during that period. Customers are consequently charged for the actual energy they consume during each time band on an ex-post basis. Possible variations to this basic option include either seasonality – where charges during the 'peak' season are higher than during the rest of the year, reflecting the greater likelihood the network will be constrained – or granular time bands – i.e. a

DNO region is broken into smaller areas with different time bands applied to reflect local network conditions.

The main advantages of volumetric ToU tariffs are that they incentivise customers to reduce peak consumption on an ongoing basis. Consumption-based charging is the current approach and so may be the easiest for customers to understand. In addition, all but small users already have a time-of-use element. Time bands and charges are set in advance and so are more transparent and predicable, enabling customers to more easily identify when they need to shift consumption patterns. Customers only face charges relating to their actual consumption, rather than incurring costs for capacity they do not require. However, static time-of-use periods may not align with actual peak periods, resulting in inefficient network usage. Because the time bands are set in advance, they may over-reward flexibility when not required and under-reward it during actual constrained periods. They may also create a 'cliff edge' where consumption shifts in response to price signals, but results in a new peak period being created at the time the peak unit rate ends – could particularly be a challenge when smart devices are ubiquitous and can be automated to respond to price signals.

Existing studies can shed light on two critical aspects of volumetric ToU tariffs: average demand reduction in correspondence with peak periods and distributional effects (Bradley et al., 2013).

With regard to average demand reductions, a review of 163 reviewed studies shows that peak reduction levels range from next to 0 to almost 60% (Faruqui & George, 2005). Findings show that peak to off-peak price ratio of 5:1, pricing only trials, obtained a 13.8% peak reduction, whereas peak to off-peak price ratio of 10:1 pushed peak reduction to almost 16%.

ToU tariffs in Ireland reduced peak consumption by 8.8% for specifically designed price bands (Darby & McKenna, 2012). In France ToU tariffs were combined with weather inputs, with days differentiated based on price signals for peak and off-peak hours. It was estimated that for a typical household with a 1 kW average load, the Tempo tariff brought about a reduction in consumption in the region of between 15% and 45%, with customers saving 10% on average on their electricity bills (Torriti et al., 2010). In Italy, ToU tariffs have gradually been applied to Italian residential electricity users since the year 2010. The first pilot of ToU (dual tariff) involved 4 million end users. Lower tariffs are applied to weekends and to weekdays from 7.00 PM to 8.00 AM. The two tariffs (initially set at 0.09982 cent/kwh and 0.07078 cent/kwh for peak and off-peak, respectively) were originally designed to yield savings for the end user whose consumption is concentrated for more than 66% during the lower tariff periods. A 2012 study finds a modest level of average peak reduction. When there is significant demand shifting, this did not necessarily follow a price-related logic. For example, in the Northern Italy data a third peak emerged in the middle of the afternoon bringing about lower consumption before the change in tariffs at 7.00 PM (Torriti, 2012).

In the United Kingdom, despite the fact that about 5.5 million customers make use of multi-rate energy tariffs and three million specifically on ToU tariffs – the most popular being called 'Economy 7' – evidence on their effects is scarce as data is not available or published (Buryk et al., 2015). Some of demand reductions at peak are inevitably intertwined with the performance of electric storage heating (Barton et al., 2013). This was integrated as heating in several residential buildings (especially in council housing blocks) as part of the nuclear power programme from the 1960s with the requirement for nuclear generators to operate continuously, giving rise to low baseload and off-peak prices (Torriti, 2015). However, the Low Carbon London trial found average shifts in energy demand of around 4.2% (UK Power Networks, 2014).

With regard to distributional effects, in principle, ToU tariffs may generate concerns about how their implementation may affect different socio-demographic groups. In particular, those who consume electricity at more expensive peak periods, and who are unable to change their consumption patterns, could end up paying significantly more. For example, those who consume electricity during the more expensive peak periods and those who are unable to change their consumption patterns (e.g. elderly pensioners and heating) could end up paying significantly more.

Understanding the distributional effects of ToU tariffs becomes vital to ensuring affordability of energy bills, at the same time as making demand more flexible. Whilst there is significant research on fuel poverty in relation to aggregate level of consumption of electricity, little is known about the effects of dynamic tariffs on different socio-demographic groups.

In the United Kingdom, the energy regulator (Ofgem) commissioned work to understand potential distributional impacts of ToU on different socio-demographic groups of consumers. Using half-hourly electricity demand data from the Energy Demand Research Project and the Low Carbon London trials, a study by Cambridge Economic Policy Associates (2017) calculated the bills that result by households based on "behavioural" estimates in response to the tariff they have selected. Findings show that there are households in all groups that would be worse off under ToU tariffs to some modest degree. According to this study, on average all socio-demographic groups, except some of the most affluent groups, would save on their annual bills. The limitations of this study relate to the use of aggregate socio-economic segmentation, which does not allow disaggregated analysis or clustering according to sets of socio-demographic variables; the uptake of tariffs based on a review of survey data rather than trials; and the use of behavioural responses from the Low Carbon London trial which is not nationally representative. The Centre for Sustainable Energy (2014) carried out analysis of the distributional effects of ToU tariffs. This showed that most consumers would see relatively small changes in bills, but some could see increases in bills of up to 20%. However, this study did not consider the effects on different socio-demographic groups and did not consider whether households indeed had flexibility in terms of how their schedules are structured. A review of US trials by Faruqui and Sergici (2013) found that low-income groups are associated with lower peak reduction than other groups. A review of pilots by Stromback et al. (2011) found that age, income, education, household size, load profile and environmental factors, such as house type, house size, and house age, are rarely captured by studies. A review by Frontier Economics and Sustainability First (2012) found no studies specifically collected information on vulnerable groups as defined by the government's Fuel Poverty Strategy (people with a long-term illness, families with children, disabled people and the elderly). In terms of uptake of the tariffs, stated preference studies on ToU tariffs show that age, gender, housing tenure (social renter, private renter, homeowner), employment status, education, income and social grade do not reveal any significant variation across groups (Nicolson et al., 2017). It can be concluded that evidence on the relationship between socio-demographic variables and response to ToU tariff is not conclusive (Hledik et al., 2017).

Tariffs based on actual capacity

The principle behind tariffs based on actual capacity is that customers will be charged for their actual capacity with charges that could vary at different times of the day. This means that consumers are charged based on their actual maximum capacity. For customers with smart meters, this will be their maximum average demand over a half hour on the network, measured ex-post. As a result, consumers might only face a charge for their maximum actual capacity during a specific peak period – e.g. during the traditional evening peak – that reflects when the network is likely to be constrained. A variation of flat actual capacity charging would lead to customers facing different rates for capacity measured during different time bands. The maximum actual capacity measurement would be reset at specific intervals (e.g. monthly, quarterly, annually). One advantage of this type of tariffs is that if network costs are driven by system peak capacity, rather than consumption, customers will only face charges for the maximum contribution they have made during the system peak. This could also apply to different time bands if using a time-of-use approach. This approach to tariffs implies that consumers only face charges relating to their actual consumption, rather than incurring costs for capacity they do not actually use. For small users in countries where capacity charges have not been applied before, it may be more difficult to understand the concept of capacity than volumetric charging, as it would be a new concept. At the same time, as part of these tariffs, consumers do not have an incentive to manage their usage on an ongoing basis, provided they stay below the maximum power demand recorded in the relevant period. This may make diversity assumptions more difficult to apply. There may be limited differences in customer response between this and volumetric ToU tariffs, where actual capacity charges also vary between time bands as capacity needs are measured at different times of the day.

Tariffs based on agreed capacity

According to tariffs based on agreed capacity, consumers are charged based on an agreed level of capacity. The level of agreed capacity could vary at different times of the day, or part of it could be curtailable. Each consumer – often via retailers – would agree *ex ante* with their DNO the maximum capacity they require on the network. Each consumer would pay a £/kW charge based on the level of agreed capacity. When consumers exceed their agreed capacity, they could be subject to curtailment or face exceedance charges. For small users, it may be more proportionate to determine their capacity requirements either as choices from a set of capacity levels or deemed by the supplier or DNO on the customer's behalf.

Because a core capacity consists of an essential amount of electrical power (expressed in kW) at the end-user level, by either limiting or charging less for core capacity, consumers might in principle benefit from lower bills. This is because paying for unutilised power capacity is expensive and arguably unfair for those who never reach high power demand. However, should core capacity be set a level which is too low, a consumer might have to pay extra for demand above such level or even face disconnection, depending on the type of core capacity arrangement in place. Different countries have dealt with the core capacity issue following a variety of approaches and much can be learned by examining such experiences. In Italy, Spain and Portugal the core capacity is related to capacity limits. This means that end users cannot exceed the capacity limit, or they would experience disconnection at the meter level. In these three countries, the capacity limit is defined according to the customer's needs. Each consumer must choose their capacity limit according to their needs, taking into account that if they request a higher capacity, the electricity bill will be higher and if they request less than they need, the meter will switch electricity off through a power breaker (Snodin et al., 2019a). In Italy, the so-called 'capacity limit' is the maximum power level made through *ex ante* agreement with the retailer. It is set when a customer contracts for supply and is based on information about the type and number of electrical appliances normally used and, for domestic customers, taking into account historical monthly maximum demand. The total number of residential users with capacity limits in Italy is 28 million, and 90% of residential customers have a capacity limit of 3kW. In addition, capacity limits are also applied to non-residential users up to 15 kW. Since 2017, customers are allowed to better adapt their limits from 0.5kW to 6kW of capacity limit in steps of 0.5 and in steps of 1kW from 6 to 10kW (ARERA, 2017). The Italian experience suggests that electricity suppliers should be obliged to disclose information on peak power on a monthly and yearly basis. In Italy this has to be included in bills because the first generation of smart meters did not supply such information. In technical terms, maximum load is taken from smart meter measurements of electricity withdrawn every 15 minutes. With regard to the main advantages of tariffs based on agreed capacity, if network costs are driven by system peak capacity, customers face charges that reflect their maximum possible contribution to the system peak. As the network is sized for peak capacity, this type of tariffs may provide greater certainty of customer usage for network companies. The cost for the latter consists of additional monitoring and enforcement applied for exceedance. For example, in Italy for any capacity limit agreed with the supplier, the consumer can overtake that limit by 10% (ARERA, 2016). For example, for a capacity limit of 3kW, the consumer may use power at 3.3kW for unlimited time. It is possible to demand on average up to 27% more than the available power (i.e. 4.2kW), calculated as average value over a time interval of two minutes. Overtaking this value means that the controller interrupts supply. As an additional advantage, tariffs based on agreed capacity could be used to incentivise different access rights choice and could reduce the price volatility of charges. On a less positive note, these tariffs could provide limited incentive for customers to reduce their peak usage below their agreed capacity, as they would still have to pay for the full capacity. This approach to planning for capacity may not align with how the DNOs plan their networks, which could mean charges are not cost-reflective. It might also be difficult to agree capacities with smaller users and identify changes in their usage over time. For instance, when people buy an electric vehicles or heat pumps, they might need to change their agreed capacity level.

Real-time pricing

Wholesale prices of electricity can oscillate significantly between low demand night-time hours and high demand evening peaks. Economists have long debated over the allocative inefficiencies of flat pricing (Holland & Mansur, 2006). The literature on real-time pricing emphasises positive net welfare effects for both business and policy purposes (Allcott, 2011). As part of real-time pricing customers are charged for their actual usage based on rates that could vary close to real-time. The time interval according to which the charging takes place could vary depending on the settlement period in the wholesale electricity market. For instance, Ofgem at the time of writing this book was in the process of reforming the settlement period to half hour. Hence, the rates could change for each half hour period and would be determined based on real-time local network conditions. Customers would be notified of the price a short period in advance. According to energy economists, real-time pricing provides the most efficient signals about when the network is constrained with the ability to vary the strength of the signal in near real-time to reflect the actual cost of managing constraints. This approach is designed to incentivise flexible services to reduce usage in response to sharp price signals. For tariffs to be based on real-time information, a high level of network monitoring and forecasting would be required. This is because network conditions need to be understood fully so that the price can be set for the next settlement period, e.g. the following hour or half hour. Real-time pricing would require the prices to be administratively set. With regard to price elasticity, consumers who are locked-in because of scarce time availability and lower income might not be able to rip the benefits of such pricing approach. What is more, large portions of consumers might not have sufficient flexibility capital in order to benefit from real-time pricing (Powells & Fell, 2019).

Critical peak pricing

Critical peak pricing consists of tariffs based on critical peaks for certain hours on event days. The event days may be advertised by the utility a day in advance, based on their forecast of a particularly high demand. In the low-carbon future, a higher number of peak events is likely to occur because of unusually high demand with the connection of electric vehicles and electric heat pumps (Foxon, 2013), and also due to higher levels of supply from intermittent sources of energy. Energy suppliers have incentives to set higher prices at peak times in order to shift demand to off-peak times and efficiently reduce capacity costs. Smart meters will provide them with the opportunity to increase the number of end users who could be notified about the very significant change in price. Through critical peak pricing energy suppliers exploit prior knowledge about when demand peaks occur. Customers face a high charge during a critical peak period and low or no forward-looking charge for the rest of the year. Typically, the rate is set in advance of the charging year, and the critical peak periods are determined dynamically and notified a short period in advance (e.g. a day ahead). The rate could be charged on the basis of capacity or consumption. The number of critical peak periods could be set in advance or vary year-by-year, depending on network conditions. In the future, critical peaks are likely to occur only a limited amount of times per year, somewhere in the region of 10–15 events per year, but may push prices up to 3–10 times as much as default tariffs (Energy Insights, 2012). In addition to the number of periods being potentially variable, it could also be possible for the length of the critical peak period to vary, rather than being fixed for each period during a charging year. In principle, critical peak pricing is a more economically efficient than a static charging option, as peak charging periods more accurately reflect real-time network conditions. The limited number of times per year when a customer needs to shift consumption in response to the charging signals might play favourably to different types of consumers. This type of tariff creates an incentive for flexible services to reduce usage. However, some feasibility issues might take place because of the level of network monitoring and forecasting required to determine when a critical peak period is expected. For example, the critical peak might prove difficult to predict and may require significant customer investment or acquisition of third party services. Also, there is the possibility of 'cliff edges' happening. This is at times when consumption shifts in response to price signals, resulting in the creation of a new peak period after the notified peak period ends.

Critical peak rebates

Through critical peak rebates customers receive a rebate for reducing usage during a period, which is labelled as 'peak'. This approach is similar to critical peak pricing, but consumers are not penalised through higher prices during the critical peak time. Instead, they receive a rebate for reducing their capacity or consumption for the period of the critical peak. The key difference is that it would be necessary for a baseline to be agreed with the DNO, in order to be possible to determine if a user has taken actions that should result in payment of a rebate. The charge for the baseline would need to be sufficient to avoid creating an incentive for customers to set the baseline higher than necessary in order to ensure they receive a rebate. Indeed, issues of setting the baseline and gaming are issues which are typical of demand-side response for commercial and industrial end users (Curtis et al., 2018). A study by The Brattle Group and University College London (2017) suggests that critical peak rebates may drive a greater response from customers than a critical peak charge. This is because where customers are on a pass-through tariff, a rebate is likely to be more acceptable than a sharp penalty during a critical peak period. However, the issue of establishing a baseline in order to identify whether a customer has taken the desired action is not likely to find easy solutions. Like other dynamic pricing options combined with smart metering, the level of network monitoring and forecasting is high in order to establish whether a critical peak period is expected.

Block pricing

Increasing block pricing is an approach which prices energy in volume-based 'blocks'. Additional volume attracts higher prices. This approach in combination with smart meters could provide a core level of affordable energy. In the telecommunications sector, there are similar tariff structures with cross-subsidies among call categories, where high-rate long-distance calls are used to pay for low line-rentals charges to reduce the bill paid by lower income consumers.

There is conflicting evidence on whether structured volume-based charging could offer a viable alternative to capacity-based charging. Increasing block pricing has precedent elsewhere and could offer a useful alternative to other forms of charging. This should be included as an option in any further consideration of capacity-based charging and investigated as to whether it is a cost reflective alternative for all parts of the United Kingdom. The literature highlights that progressive tariffs do not bring about higher efficiency, and consumers tend to respond to average price signals (i.e. ratio between total bill price and consumption in kWh) and not to marginal prices (i.e. c€/kWh of the specific step) (Kahn & Wolak, 2013). The extent to which progressive tariffs are effective at influencing behaviour depends significantly on the ways in which the increments between steps are defined and on the price elasticity of demand (Koichiro, 2010).

The Italian experience with progressive block tariffs challenges does not show any significant behavioural change in terms of reduction to energy consumption. Knowledge of progressive tariffs by Italian residential consumers seems to be extremely low because of the absence of any systematic information campaign and the frequent changes to the structure of the steps. The lack of variation across regions puts more importance to price elasticity of demand. As a way of example, the 1,800 kWh/year threshold, on the one hand, can be too high for a single-person household and, on the other hand, can be too low for a family with several children. The fixed nature of the steps disregards actual electricity needs in relation to different climatic conditions and the availability of different fuels (e.g. fuel switching for heating purposes). The capacity limit has contributed more strongly to the limited growth in electricity consumptions since the 1970s in Italy. The split in tariff steps implies average price signals which are hardly perceived by residential consumers and hence is unlikely to influence behaviour. Most residential consumers' consumption is between 1,000 and 2,700 kWh/year, and average price varies between 16.5 and 18.7 c€/kWh (Snodin et al., 2019b).

Key messages of this book

Predecessors of the smart meter faced similar challenges in terms of demonstrating their economic case

Throughout the history of electricity meters since the last quarter of the 19th century, their costs were a very significant criterion for the development of successive versions of electricity meters. There has always been an intricate relation between the meters and units, with the technical configuration and functionalities of the former often determining the latter. This leads to further thinking, for instance, around whether in the future it will be more useful to have kilowatt-hours or kilowatts as units of measurement as the amount of power demanded at any given time and the peaks associated with it will have more significant impact on networks and infrastructure than average volumes of consumption. Smart metering is not a commodity service, but a fundamental infrastructure asset.

Assessing the costs and benefits of smart meters is a difficult task

The review of smart meter cost–benefit analyses (CBAs) in European countries points to the challenges of carrying out precise economic assessments in this area. CBAs were positive and roll-outs are therefore underway or being planned for electricity meters in 16 member states, whereas seven member states found negative CBAs for the large-scale roll-out of electricity meters.

The costs of national roll-outs vary between countries because of labour costs, but also tendering procedures leading to different values of costs even within the same country (as was the case in Romania in Chapter 4). The benefits vary greatly between the Distribution System Operators (DSOs) of countries with high levels of theft and commercial losses and DSOs of countries which do not experience this contextual problem. In several countries, the Net Present Values (i.e. the difference between benefits and costs) is next to 0. Thus far, following recommendations from the European Commission in this area, European countries have followed relatively standardised approaches in terms of categories of costs and benefits. This means that aspects of the smart grid have not been included in the economic assessments. The extent to which smart meters will facilitate the uptake of flexibility services will shape future CBAs as a whole family of new benefits will come into play, including improving distribution network investment efficiency, contributing to the management of the demand–supply balance and increasing interaction with distributed power systems.

Capping and tendering

The Romanian case study in Chapter 4 shows that the maximum unit cost was 250% higher than the minimum one. Smart meter providers charged very different prices to different DSOs based on 'perfect foresight' of how much DSOs bidded for in the tender documents. This poses questions as to how capping and competitive pricing could be in place for smart meters. Possible options for capping smart-metering price include national tender, forcing a competitive tendering approach, price capping approach, price capping approach combined with capital investment in infrastructure, rate-of-return approach and yardstick competition approach.

The use of depreciation rates associated with the new meters poses questions as to how this should be treated in CBA practice. As a general principle, only real costs and benefits, i.e. changes in real resources, should be taken into account. For instance, accounting depreciation expenses should not be taken into account, since this would double-count the capital investment that has already been accounted as a cost. The risk of double counting when considering depreciation charges is high (as happens for interest and cost of capital). For instance, a depreciation charge is intended to reflect to 'consumption' of capital, or the reduction in the value of the capital investment over a specified period, but would double count the cost of an investment if the smart-metering cost was already included in the CBA. Accounting practice is to treat asset costs as capital expenditure and to recognise depreciation as an operating cost. The usual practice in CBA is to recognise capital expenditure when it is incurred, and ignoring depreciation. This simplifies the task of ensuring that the time value of money (i.e. discounting) is properly taken account of.

With regard to approaches taking into account depreciation in CBA, when almost the entirety of costs and benefits is borne by asset owners, the CBA has several features which normally characterise a private CBA. In the

example of CBA on smart meters, it is not uncommon to consider depreciation levels of old and new meters. It then becomes possible to include a value
related to the depreciation of assets.

Cost–benefit analysis applied to smart meters has several limitations

Over the years, policy-makers have expanded the types of analytical tools
used in the appraisal and evaluation of energy policies from narrowly scoped
geophysical and ecological models, on the one hand, and purely socio-
economic-oriented tools of decision analysis, on the other hand, to highly
integrated assessment tools, such as CBA. The use of CBA for infrastructure energy programmes and policy compels a comprehensive approach
with regard to including all societal costs and benefits. When it comes to
smart meters, there are significant limitations on the extent to which CBA
can capture all types of benefits. Several assumptions are made in smart-
metering calculations, and these tend to be rather limited in terms of the
extent to which the smart grid will be developed, electric vehicles will interact with smart meters and the presence of home battery storage. In essence,
CBAs tend to be anchored to extensions of representations of the present.

For instance, the most recent CBA by BEIS (2019) identifies areas which
cannot be easily captured and monetised. These include improved budget-
ing and peace-of-mind, particularly for prepayment consumers; increased
convenience and reassurance from avoided need to read meters or allow
meter readers into the property; and the possibility for 'comfort taking',
which is defined as enabling consumers to have the confidence to spend
more to achieve appropriate levels of comfort owing to better understanding of energy spending through greater access to data and real-time energy
use. The same CBA expects that smart meters will enhance the ability of
consumers in vulnerable groups to identify cheaper tariffs that enable
appropriate heating of homes to be afforded.

The economic case for smart meters based on net conservation effects is doomed

Before reading this book, it would be clear to anyone that smart meters alone
cannot deliver reductions in energy demand. This is because they are devices
through which information about electricity consumption is communicated
to users and transmitted to network operators. Less clear is the extent to
which smart meters and the information they provide could – in principle –
nudge behaviour and lead to changes in energy consumption (whether in
absolute terms or at specific times of the day through load shifting).

Recent CBAs (e.g. BEIS, 2019) make use of net conservation effects figures
of 3%. The review carried out as part of Chapter 6 shows that over the years
there has been a consistent diminution of net conservation effects. This
reduction is significant when considering that more recent studies showing

lower net conservation effects are more reliable in terms of sample size and methods.

Smart meters: which use can they have for economists and social scientists?

Whilst conceptualising smart meters is a difficult task, the discussion on behavioural and social practice theory provided a fertile ground for reflecting on their role in different domains of the social sciences. Smart meters have already been a territory for applied research on 'nudging' and the measurement of individuals' attitudes, behaviours and contexts. Smart meters also offer information, in the form of energy demand data, which can be seen as some form of monitoring of what people do, when they are at home, at work or charging their electric vehicles. Smart meters can leave traces of everyday life, especially if the data they generate is examined in trends, over seconds, minutes, hours, days, weeks, months, seasons, years and (one day) even decades. In future research smart meters could play a vital role not only in the study of individuals' behavioural change in response to information and price inputs, but also on how bundles of practices vary across time and space.

References

ACEEE. (2019). Energy as a service. *ACEEE Emerging Opportunities Series.* Retrieved from October 2019 https://aceee.org/sites/default/files/eo-energy-as-service.pdf

Allcott, H. (2011). Rethinking real-time electricity pricing. *Resource and Energy Economics, 33*(4), 820–842.

Andoni, M., Robu, V., Flynn, D., Abram, S., Geach, D., Jenkins, D., McCallum, P., & Peacock, A. (2019). Blockchain technology in the energy sector: A systematic review of challenges and opportunities. *Renewable and Sustainable Energy Reviews, 100*, 143–174.

ARERA. (2016). Relazione AIR - Riforma delle tariffe di rete e delle componenti tariffarie a copertura degli oneri generali di sistema i clienti domestici di energia elettrica. [Accessed online on 1st March 2019]. Retrieved from https://www.arera.it/it/elettricita/auc.htm

ARERA. (2017). Potenza del contatore, agevolazioni e maggiore scelta. Nuove possibilità di utilizzo dell'energia elettrica e risparmio per le famiglie italiane. [Accessed online on 1st March 2019]. Retrieved from https://www.arera.it/allegati/consumatori/17potenzacont.pdf

Barton, J., Huang, S., Infield, D., Leach, M., Ogunkunle, D., Torriti, J., & Thomson, M. (2013). The evolution of electricity demand and the role for demand side participation, in buildings and transport. *Energy Policy, 52*, 85–102.

BEIS. (2017a). 2016 UK Greenhouse gas emissions, provisional figures. [Accessed online on 1st March 2019]. Retrieved from https://assets.publishing.service.gov.uk/government/uploads/system/uploads/attachment_data/file/604408/2016_Provisional_Emissions_statistics.pdf

BEIS. (2017b). UK energy in brief 2017. [Accessed online on 1st March 2019]. Retrieved from https://assets.publishing.service.gov.uk/government/uploads/system/uploads/attachment_data/file/631146/UK_Energy_in_Brief_2017.pdf

BEIS (2019). Smart meter roll-out cost benefit analysis. [Accessed online on 1st March 2019]. Retrieved from: https://assets.publishing.service.gov.uk/government/uploads/system/uploads/attachment_data/file/831716/smart-meter-roll-out-cost-benefit-analysis-2019.pdf

Bradley, P., Leach, M., & Torriti, J. (2013). A review of the costs and benefits of demand response for electricity in the UK. *Energy Policy, 52*, 312–327.

Buryk, S., Mead, D., Mourato, S., & Torriti, J. (2015), Investigating preferences for dynamic electricity tariffs: the effect of environmental and system benefit disclosure. *Energy Policy, 80*, 190–195.

Cambridge Economic Policy Associates. (2017). Distributional impact of time of use tariffs. Technical Report 416.

Centre for Sustainable Energy. (2014). Investigating the potential impacts of Time of Use (ToU) tariffs on domestic electricity customers: Smarter markets programme. Technical Report April. [Accessed online on 1st March 2019]. Retrieved from https://www.ofgem.gov.uk/publications-and-updates/investigating-potential-impacts-time-use-tariffs-domestic-electricity-customers-smarter-markets-programme

Crosby, M., Pattanayak, P., Verma, S., & Kalyanaraman, V. (2016). Blockchain technology: Beyond bitcoin. *Applied Innovation, 2*(6–10), 71.

Curtis, M., Torriti, J. & Smith, S. T. (2018). A comparative analysis of building energy estimation methods in the context of demand response. *Energy and Buildings, 174*. 13–25.

Danieli, A. (2018). The French electricity smart meter: Reconfiguring consumers and providers. In Shove, E. & Trentmann, F. (eds.), *Infrastructures in Practice* (pp. 155–168). Abingdon: Routledge.

Darby, S. J. (2018). Smart technology in the home: Time for more clarity. *Building Research & Information, 46*(1), 140–147.

Darby, S. J., & McKenna, E. (2012). Social implications of residential demand response in cool temperate climates. *Energy Policy, 49*, 759–769.

Department of Energy. (2008). What Is efficiency-as-a-service? [Accessed online on October 2019] betterbuildingssolutioncenter.energy.gov/financing-navigator/option/efficiency-a-service.

Energy Insights. (2012). Worldwide quarterly smart meter tracker. [Accessed online on 1st March 2019]. Retrieved from http://www.idc.com/tracker/showproductinfo.jsp?prod_id=79

European Commission (2006). European smart grids technology platform: Vision and strategy for Europe's electricity networks of the future. Luxembourg: Office for Official Publications of the European Communities.

Faruqui, A., & George, S. (2005). Quantifying customer response to dynamic pricing. *The Electricity Journal, 18*(4), 53–63.

Faruqui. A., & Sergici. S. (2013). Arcturus: International evidence on dynamic pricing. *Electricity Journal, 26*(7), 55–65.

Fell, M. J., Schneiders, A., & Shipworth, D. (2019). Consumer demand for blockchain-enabled peer-to-peer electricity trading in the United Kingdom: An online survey experiment. *Energies, 12*(20), 3913.

Foxon, T. J. (2013). Transition pathways for a UK low carbon electricity future. *Energy Policy, 52*, 10–24.

Frontier Economics & Sustainability First. (2012). Domestic and SME tariff development for the Customer - Led Network Revolution. Technical Report CLNR-L006 June.

Geels, F. W. (2004). From sectoral systems of innovation to socio-technical systems: Insights about dynamics and change from sociology and institutional theory. *Research Policy, 33*(6–7), 897–920.

Gellings, C. W. (2009). *The smart grid: Enabling energy efficiency and demand response.* Lilburn: The Fairmont Press, Inc.

Gram-Hanssen, K., & Darby, S. J. (2018). Home is where the smart is? evaluating smart home research and approaches against the concept of home. *Energy Research & Social Science, 37*, 94–101.

Hanna, R., Leach, M. & Torriti, J. (2018). Microgeneration: The installer perspective. *Renewable Energy, 116*, 458–469.

Hargreaves, T., & Wilson, C., (2017). *Smart homes and their users.* Switzerland: Springer.

Hargreaves T., Wilson, C., & Hauxwell-Baldwin, R. (2018). Learning to live in a smart home. *Building Research & Information, 46*(1), pp. 127–139.

Hazas, M., & Strengers, Y. (2019). Promoting Smart Homes. In Rinkinen, J., Shove, E. & Torriti, J. (eds.), *Energy fables: Challenging ideas in the energy sector* (pp. 78–88). Abingdon: Routledge.

Hledik, R., Gorman, W., Irwin, N., Fell, M., Nicolson, M., & Huebner, G. (2017). The value of TOU Tariffs in great britain: Insights for decision-makers. Technical Report. [Accessed online on 1st March 2019]. Retrieved from https://www.citizensadvice.org.uk/Global/CitizensAdvice/Energy/The%20Value%20of%20TOU%20Tariffs%20in%20GB%20-%20Volume%20I.pdf

Holland, S. P., & Mansur, E. T. (2006). The short-run effects of time-varying prices in competitive electricity markets. *The Energy Journal, 27*, 127–155.

IRENA. (2018). Blockchain: A New Tool to Accelerate the Global Energy Transformation, *International Renewable Energy Agency.*

Kahn M. E., & Wolak F. A. (2013). Using Information to improve the effectiveness of nonlinear pricing: Evidence from a field experiment. http://web.stanford.edu/group/fwolak/cgibin/sites/default/files/files/kahn_wolak_July_2_2013.pdf

Koichiro I. (2010). Do consumers respond to marginal or average price? Evidence from nonlinear electricity pricing. University of California, http://www.economics.utoronto.ca/index.php/index/research/downloadSeminarPaper/4174

Mengelkamp, E., Gärttner, J., Rock, K., Kessler, S., Orsini, L., & Weinhardt, C. (2018). Designing microgrid energy markets: A case study: The Brooklyn Microgrid. *Applied Energy, 210*, 870–880.

Molina-Garcia, A., Fuentes, J. A., Gomez-Lazaro, E., Bonastre, A., Campelo, J. C., & Serrano, J. J. (2007). Application of wireless sensor network to direct load control in residential areas, *IEEE International Symposium on Industrial Electronics*, Caixanova-Vigo, Spain, 1974–1979.

Nicolson, M., Huebner, G., & Shipworth, D. (2017). Are consumers willing to switch to smart time of use electricity tariffs? The importance of loss-aversion and electric vehicle ownership. *Energy Research and Social Science, 23*, 82–96.

Nyborg, S., & Røpke, I. (2011). Energy impacts of the smart home – Conflicting visions. Paper presented at the ECEEE 2011 Summer Study (6–11 June 2011), Toulon, France.

Paetz, A. G., Dütschke, E., & Fichtner, W. (2012). Smart homes as a means to sustainable energy consumption: A study of consumer perceptions. *Journal of Consumer Policy, 35*(1), 23–41.

Powells, G., & Fell, M. J. (2019). Flexibility capital and flexibility justice in smart energy systems. *Energy Research & Social Science, 54*, 56–59.

Snodin, H., Torriti, J., Anderson, M., & Jones, E., (2019a). Core capacity: Interrogation of SAVE data & social science literature review. Report. SSEN

Snodin, H., Torriti, J., & Yunusov, T., (2019b). Consumer network access, core capacity. Report. Citizens Advice.

Strengers, Y., & Nicholls, L. (2017). Convenience and energy consumption in the smart home of the future: Industry visions from Australia and beyond. *Energy Research & Social Science, 32*, 86–93.

Stromback, J., Dromacque, C., & Yassin, M. (2011). The potential of smart meter enabled programs to increase energy and systems effciency: A mass pilot comparison. [Accessed online on 1st March 2019]. Retrieved from URL: papers2:// publication/uuid/45A4910F-C10F-4873-B2662311EF22FD5E

Sustainability First and Centre for Sustainable Energy. (2019). Annex to PIAG final report: Summary of PIAG project papers – Phase 1. [Accessed online on 1st March 2019]. Retrieved from https://www.smartenergydatapiag.org.uk/

The Brattle Group and University College London. (2017). The value of TOU tariffs in great Britain: Insights for decision-makers. [Accessed online on 1st March 2019]. Retrieved from https://www.citizensadvice.org.uk/Global/CitizensAdvice/ Energy/The%20Value%20of%20TOU%20Tariffs%20in%20GB%20-%20 Volume%20I.pdf

Torriti, J. (2012) Price-based demand side management: Assessing the impacts of time-of-use tariffs on residential electricity demand and peak shifting in Northern Italy. *Energy, 44*(1), 576–583.

Torriti, J. (2014). People or machines? Assessing the impacts of smart meters and load controllers in Italian office spaces. *Energy for Sustainable Development, 20*, 86–91.

Torriti, J. (2015). *Peak energy demand and demand side response.* Routledge Explorations in Environmental Studies. Routledge, Abingdon.

Torriti, J. (2017). The risk of residential peak electricity demand: A comparison of five European countries. *Energies, 10*(3), 385–399

Torriti, J., Hassan, M. G., & Leach, M. (2010). Demand response experience in Europe: Policies, programmes and implementation. *Energy, 35*(4), 1575–1583.

UK Power Networks. (2014). UK Power networks, low carbon london, project progress report, June 2014. [Accessed online on 1st March 2019]. Retrieved from http://innovation.ukpowernetworks.co.uk/innovation/en/Projects/tier-2-projects/ Low-Carbon-London-%28LCL%29/Project-Documents/LCL+June+2014+PPR+ v1.1+PXM+250614+%28changes+accepted%29.pdf

Van Dam, S., Bakker, C., & van Hal, J. (2010). Home energy monitors: Impact over the medium-term. *Building Research & Information, 38*(5), 458–469.

World Economic Forum. (2017). *The future of electricity: New technologies transforming the grid edge.* Geneva: World Economic Forum.

Williams, E. Matthews, S., & Brady, T. (2006). Use of a computer-based system to measure and manage energy consumption in the home, in *Proceedings of the 2006 IEEE International Symposium on Electronics and the Environment* (167–172), Vol. 0, Scottsdale, USA.

Appendix – policy options in regulatory impact assessments on smart meter roll-outs

	Option 1	Option 2	Option 3	Option 4	Option 5
Australia	Option 1: A continuation of the installation of manually read accumulation meters on a new and replacement basis (noting that most distributors were installing electronic, twin register accumulation meters that were capable of separately recording peak and off-peak usage, with a separate peak rate for customers with electric storage hot water systems).	Option 2: The gradual uptake of new metering technology (either manually read or remotely read interval meters) by distributors (or retailers) as the market for these more sophisticated meters evolves and they become cost-comparable to accumulation meters.			
Belgium (Brussels)	Basic: Factureren op maandbasis; meetwaarden per dag en 1 keer per maand doorsturen; aan- afsluiten en verhuizen op aanvraag; meest basic smart meter voldoet; PLC communicatie als minimum; overal uitrollen.	Moderate: Basic plus: energiebesparing gerealiseerd dmv terugkoppeling via internet; 15 minuten intervalwaarden 1 keer per dag doorsturen voor energiebesparingen waardoor PLC krap wordt dus UMTS.	Advanced: Sustainable plus: Ook spanning en frequentie meten voor om. geavanceerde fraudedetectie; 15 minuten intervalwaarden 1 keer per uur doorsturen; geen kosten meegenomen voor publieke laadpalen (in laadpaal zit slimme meter ingebakken). UMTS voldoet nog steeds, extra bandbreedte van Wimax niet perse nodig. (*EU option*)	Full: Smart grid plus: Kortere responstijden: kwartierwaarde quasi realtime doorsturen en communicatie met IHD of equivalant via bestaande internetaansluiting; Wimax beste keuze naar de toekomst toe, anticiperend dat tegen uitrol wel licenties beschikbaar zijn.	

(Continued)

	Option 1	Option 2	Option 3	Option 4	Option 5
Belgium (Flanders)	A uniform roll-out over 5 years.	Segmented roll-out over 6 years. This starts with larger commercial customers, moving over time to residential customers. This option is considered in this section as most closely approximating the expectations of Recommendation 2012/148/EU.			
Belgium (Wallonia)	Full Roll-out: 80% of customers fitted with electricity and gas meters by 2020.	Smart-meter friendly: selective roll-out to particular customer segments, including those with a bad payment record, new installations, replacements and those customers requesting a smart meter and paying for the installation.			

Country					
Brazil	No intervention: The electronic meter registers consumption of active energy only and will be installed in all the low voltage consumer units (68 millions in Brazil) along 25 years uniformly, equivalent to the regulatory life-time of the current conventional electromechanical meter.	Basic: In addition to the features of the meter in Option 1, reactive energy and demand are functionalities of the meter in this case, which will be implemented in all low voltage consumer units along ten years, 10% per year.	Time of use tariff and continuity indexes: In this case, two sub-options will be created: (1) all 68 million consumers receive the meter, and (2) only those consumers who request the time of use tariff will have the meter installed in your unit; it will be assumed that 10% of the consumers request the new tariff; in both sub-options, the meters roll-out would occur uniformly in 10 years.	Communication: Two sub-options will be assessed, considering the meters roll-out finishing in ten years and covering all consumers and 10% of them.	
Czech Republic	Blanket option: Introduction of smart metering according to EU Directive.	Basic option: A variant of the status quo.			
Germany	EU option: Assesses that at least 80% of all metering points must be equipped with a smart-metering system within a period of ten years.	Continuity option: Assesses the legislative approach with the current mandatory installation in accordance with German Law	Continuity plus option: Combination of forward-looking intelligent meters in addition to smart-metering systems. The intelligent meters must offer the possibility of integration into a **BSI Protection Profile**-compliant communication system.	Roll-out option: The roll-out option provides for mandatory installations for old EEG/CHP plants as well as those with a power input lower than 7 kW down to a negligibility limit of 250 Watt.	Roll-out plus option: Same as roll-out option with increased roll-out quota to facilitate further economies of scale.

(Continued)

	Option 1	Option 2	Option 3	Option 4	Option 5
Great Britain	Mandate new and replacement *LATER IMPACT ASSESSMENT*: Full Establishment: Roll-out commences when central communications systems are in place.	Voluntary roll-out *LATER IMPACT ASSESSMENT*: Staged Implementation: Roll-out begins in advance of the full establishment of the central communications system.			
Hungary	Option1: This involves a distributor-based roll-out of smart meters resulting in 80% coverage by 2021. *(EU option)*	Option 2: A joint roll-out by all DSOs, in which the companies concerned jointly develop the system that processes and stores the data provided by smart meters, and make them available to the utilities and authorities for further use. Under this policy option, there is a necessity to create a new entity known as the Smart Meter Operator responsible for data collation and transfer.	Option 3: A variation of option 2, with MAVIR the Transmission System Operator taking over the functions of the Smart Meter Operator.	Option 4: A variation of option 3 in which demand management functions are included in the smart meter set up (remote operations, etc.).	

six smart-metering roll-out options with different volumes of metering installation.

	Advanced functionality		Multi-metering
Latvia			
Lithuania			
New Zealand	Base-case		
	Option 1: Including HAN interfaces only, at the time of rolling out smart meters	Option 2: Including HAN interfaces only, retrofitted as consumers take up smart appliances	Option 3: Including HAN interfaces and IHDs, at the time of rolling out smart meters.
Portugal	Roll-out 1: Starting in 2016, aiming for 80% roll-out by 2020 and 100% by 2022.	Roll-out 2: Starting in 2014, aiming for 80% roll-out by 2020 and 100% by 2022.	
Slovak republic (both options have a 23% smart-metering roll-out cap)	Progressive option: 70% of the smart meters covered will be installed during the first 4 years and 100% of the planned target will be fitted with smart meters after 8 years, in 2020.	Linear approach: An even implementation profile for smart meters over the period to 2020.	
Slovenia	Roll-out 1 option: Deployment starting in 2015 with a target of 80% by 2020. *(EU option)*	Roll-out 2 option: Starting in 2015 with a target of 80% by 2025. The remaining 20% conventional meters are replaced naturally at the end of their lifetime or due to failure.	Roll-out 3 option: Starting in 2015 with a target of 100% by 2025. Annual number of smart meters is linear.
	Roll-out 4 option: Starting in 2015 with a target of 80% by 2030. Annual number of smart meters is linear.	Roll-out 5 option: Starting in 2015 natural replacement of all conventional meters with smart meters.	

Index

For Product Safety Concerns and Information please contact our EU
representative GPSR@taylorandfrancis.com
Taylor & Francis Verlag GmbH, Kaufingerstraße 24, 80331 München, Germany